Sounds of War and Peace

Eastern European Studies in Musicology

Edited by Maciej Gołąb

Volume 10

PETER LANG

Renata Tańczuk / Sławomir Wieczorek (eds.)

Sounds of War and Peace

Soundscapes of European Cities in 1945

PETER LANG

Bibliographic Information published by the Deutsche Nationalbibliothek
The Deutsche Nationalbibliothek lists this publication in the Deutsche Nationalbibliografie; detailed bibliographic data is available in the internet at http://dnb.d-nb.de.

Library of Congress Cataloging-in-Publication Data
A CIP catalog record for this book has been applied for at the Library of Congress.

This publication was financially supported by the Faculty of Historical and Pedagogical Sciences at the University of Wrocław, Poland.

ISSN 2193-8342
ISBN 978-3-631-75336-1 (Print)
E-ISBN 978-3-631-75346-0 (E-PDF)
E-ISBN 978-3-631-75347-7 (EPUB)
E-ISBN 978-3-631-75348-4 (MOBI)
DOI 10.3726/b13870

Peter Lang – Berlin · Bern · Bruxelles · New York · Oxford · Warszawa · Wien

This publication has been peer reviewed.

www.peterlang.com

Renata Tańczuk / Sławomir Wieczorek (eds.)

Sounds of War and Peace

Soundscapes of European Cities in 1945

PETER LANG

Bibliographic Information published by the Deutsche Nationalbibliothek
The Deutsche Nationalbibliothek lists this publication in the Deutsche Nationalbibliografie; detailed bibliographic data is available in the internet at http://dnb.d-nb.de.

Library of Congress Cataloging-in-Publication Data
A CIP catalog record for this book has been applied for at the Library of Congress.

This publication was financially supported by the Faculty of Historical and Pedagogical Sciences at the University of Wrocław, Poland.

ISSN 2193-8342
ISBN 978-3-631-75336-1 (Print)
E-ISBN 978-3-631-75346-0 (E-PDF)
E-ISBN 978-3-631-75347-7 (EPUB)
E-ISBN 978-3-631-75348-4 (MOBI)
DOI 10.3726/b13870

Peter Lang – Berlin · Bern · Bruxelles · New York · Oxford · Warszawa · Wien

This publication has been peer reviewed.

www.peterlang.com

Contents

Introduction

World War II radically altered the image of many European cities. Some, like Warsaw and Dresden, were almost completely destroyed, and became symbols both of wartime barbarism and of the recuperative power of their respective nations. Others, like Wrocław (previously Breslau) and Lviv (previously Lwów-Lvov), were assigned by the signatories of international treaties to a different state, and consequently gained new populations. For many inhabitants of European cities, 1945 was a year of hope and of a return to normality; for others, however, it was the year when they were forced to leave the place they called home. These dramatic changes marked people's lives. They also affected the sensual experience of city life, as soundscapes underwent radical transformations.

We asked researchers representing various disciplines and academic centres about the specific qualities of European city soundscapes in that watershed year when war came to an end, the reception of those soundscapes and their representation in autobiographical texts and in art. This topic is important to us for both scientific and personal reasons. We live in Wrocław – a city which until May 1945 was the German Breslau. Toward the end of World War II, this city was converted into a fortress, the defence of which cost the lives of many thousands of its inhabitants and led to destruction on an enormous scale. The Germans – who had lived in Breslau for many generations – were displaced and exiled. Their place was taken by our grandparents and relatives, among others, whose task it became to rebuild and to grow accustomed to this unfamiliar, "alien" city. Wrocław's unique history of transformation has inspired us – as researchers specialising in contemporary and historical soundscapes, in cultural phenomena and sound studies[1] – to address the issues mentioned above.

---

1 We are the staff of the Soundscape Research Studio founded by Robert Losiak at the Institute of Cultural Studies of the University of Wrocław in 2009. The Studio conducts research in the areas of soundscape studies and sound studies, as well as engages in education. The Studio's most important project thus far has been a multifaceted study of the soundscapes of contemporary Wrocław, the results of which have been described in a monographic publication (*Audiosfera Wrocławia* [The sounds of Wrocław], ed. Robert Losiak and Renata Tańczuk (Wrocław: Wydawnictwo Uniwersytetu Wrocławskiego, 2014), 402 pp.), and an online soundmap of the city. The Studio also publishes the periodical *Soundscape. Concepts – Research – Practice*. More information concerning the Studio's activities can be found on its website: http://pracownia.audiosfery.uni.wroc.pl.

A comprehensive study of urban soundscapes in 1945, their reception and representation, ought to cover such areas as the following: the soundscape of air raids and bombings; silence and noise in the sound environment of ruins and empty spaces; "attentive listening" in cities fraught with danger; sound and trauma; musical creativity; the sonic aspect of Victory Day celebrations; the de-urbanisation and rusticalisation of the soundscape of destroyed cities; the sound environment of post-war reconstruction; the sonic indications of a return to normality; the constant and changing sounds of propaganda – from Nazi to Communist propaganda; sound technologies (radio, broadcasting centre, street loudspeakers); the transformations of national acoustic communities; the adaptation of unfamiliar urban spaces through sounds; the ways in which the soundscape of 1945 was represented in literature, autobiographies, feature films, documentaries, exhibitions and musical compositions. Some of these topics have already been taken up by researchers working mainly on representations of the soundscape of World War II;[2] it is to their analyses that the authors of the texts contained in the present publication refer. Our focus, however, is on the urban soundscape in 1945, conceived not as a definite moment on the axis of time, but as a symbolic turning point, a great watershed period of transition – between the end of the war and the beginning of peace.

The papers collected in this publication are studies on the soundscapes of Amsterdam, Dortmund, Lvov/Lviv, Breslau/Wrocław and Warsaw. For the purpose of describing those soundscapes and analysing their representation and reception, the authors have made use of different types of sources: autobiographical texts (Annelies Jacobs, Zoriana Rybchynska, Renata Tańczuk),

---

2 In this context, one should mention, first and foremost, the following: Carolyn Birdsall, *Nazi Soundscape. Sound, Technology and Urban Space in Germany, 1933–1945* (Amsterdam: Amsterdam University Press, 2012); Robert Maier (ed.), *Akustisches Gedächtnis und Zweiter Weltkrieg* (Göttingen: V&Runipress, 2011); selected texts dealing with the years 1933–1949 from the collection G. Paul and R. Schock (ed.), *Sound der Zeit: Geräusche, Töne, Stimmen – 1889 bis heute* (Göttingen, Wallstein, 2014); publications by Annelies Jacobs, 'The silence of Amsterdam before and during World War II: Ecology, semiotics and politics of urban sound', in Daniel Morat (ed.), *Sounds of Modern History. Auditory Cultures in 19th- and 20th-Century Europe* (New York: Berghahn, 2014), 305–325; Gabriel Moshenska, 'Moaning Minnie and the Doodlebugs. Soundscape of air warfare in Second World War Britain', in Nicholas J. Saunders and Paul Cornish (eds), *Modern Conflict and the Senses* (London: Routledge, 2017), 106–122; Joy Damousi, 'Sounds and silence of war. Dresden and Paris during World War II', in Joy Damousi and Paula Hamilton (eds), *A Cultural History of Sound, Memory and the Senses* (London: Routledge, 2017), 123–142.

interviews (Katarzyna Naliwajek-Mazurek, Uta C. Schmidt), literary texts (Jadwiga Zimpel), archive documents and press articles (Kamila Staśko-Mazur). The various papers in this book share an idiographic approach to the subject and inspiration from Raymond Murray Schafer's concept of the soundscape, as critically interpreted. The authors have also applied and developed such research tools as auditory topoi and narrative strategies created by a research team headed by Karin Bijsterveld.[3]

The studies by Annelies Jacobs and Zoriana Rybchynska deal with the sounds of wartime and post-war Amsterdam and Lvov/Lviv, as described in the diaries of their inhabitants. Uta C. Schmidt, meanwhile, focuses on the soundscapes of the Ruhr and Dortmund, ranging from the sound of sirens to the music of jazz improvisations. She indicates the ideological and political meanings of jazz in the post-war period. She also demonstrates that the sonic experience related to nocturnal air raids helped to form a new "imagined community" of victims and stresses the importance of the sounds of rebuilt industry for shaping the Ruhr's identity.

Different aspects of Warsaw's soundscape and its reception are discussed by Katarzyna Naliwajek-Mazurek, Kamila Staśko-Mazur and Jadwiga Zimpel. Naliwajek-Mazurek focuses on the memories of musicians, composers and musicologists, some of whom preserved the wartime sounds of the city in their minds and returned to them later in their works. The author addresses the sonic dimension of traumatic experience and its artistic representations and transformations. The question of trauma returns in the paper by Jadwiga Zimpel, who concentrates on silence as a feature of the post-war landscape and as a category important to the study of auditory experiences during that period. She draws on the works of one of Poland's most important writers, Miron Białoszewski, a witness to the horrors of World War II and to the post-war revival. Zimpel explores literature as aural memory and as an archive of lost sounds. Staśko-Mazur describes the complex process of the rebuilding of the radio structures and network in the ruined Warsaw of 1945, as well as the social and political significance of radio during the post-war period.

The transformation of Wrocław's soundscape is the subject of four papers. Karolina Jara discusses the aborted Nazi project for rebuilding the centre of Breslau and speculates how it would have changed the sound environment. She also addresses the influence of soundscape on new urban development plans, highlighting the ideological nature of the project, inspired by the 12th Festival (Sängerbundesfest)

3 Karin Bijsterveld (ed.), *Soundscapes of the Urban Past. Staged Sound as Media Cultural Heritage* (Bielefeld: Transcript, 2013), 11–28 and 31–66.

of the German Choral Society (Deutsche Sängerbund), held in Breslau in 1937 and used by the National Socialists for the purpose of appropriating public space. This foreshadowed in a way the tragic fate of Breslau, the post-war life of which, determined by the city's new inhabitants, would be marked by other ideological-political campaigns undertaken by the new authorities. Andrzej Dębski records artistic events in Breslau/Wrocław during the twelve dramatic months of 1945: the work of cinemas, theatres and the opera. Those events significantly contributed to the city's soundscape, providing entertainment and serving the purposes of propaganda. During the war, they reminded people of better days; after the war, they testified to a return to the routines of normal life. Renata Tańczuk and Sławomir Wieczorek explore autobiographical accounts in order to capture the ways in which Wrocław's first Polish settlers perceived and experienced the city's soundscape. They reconstruct the auditory dimension of the complex process of acclimatising to an unfamiliar city and adapting to its culturally alien space.

The group of texts devoted to the city of Wrocław is complemented by two articles by Dorota Błaszczak and Daniel Brożek, who present their own original artistic projects – time machines that transport the audience back to the soundscapes of Wrocław's past. Błaszczak's installation made use of radio archive recordings, whereas Brożek created an imaginary soundscape of the post-war cinema. The two projects differed in their strategies for reconstructing past soundscapes. They also illustrated the possibility of representing past soundscapes in contemporary artistic practice.

War is inevitably associated with traumatic experiences, frequently concealed behind silence and also associated with specific sound-related sensations. Dorian Lange goes beyond purely historical explorations of European urban soundscapes and tries to apply the results of trauma studies to his research into historical soundscapes. He points out that silence and a refusal to articulate one's traumatic experiences are common reactions; he also indicates representations of traumatic soundscapes in art.

The papers collected here demonstrate the range of possible approaches to the study of historical soundscapes, as well as the wealth of available sources. They also offer insight into the sensory, embodied experience of the city. Studies on the soundscapes of different cities also provide interesting comparative material, which should contribute to a better understanding of the year 1945, which was crucial to the formation of contemporary Europe.

Renata Tańczuk
Sławomir Wieczorek

Annelies Jacobs

Faculty of Arts and Social Sciences, Maastricht University

Barking and blaring: City sounds in wartime[1]

Abstract: This chapter deals with sound as part of everyday life in Amsterdam during the Second World War and, by drawing on diaries, investigates how its citizens reacted to new sounds, altered meanings of sounds and the disappearance of familiar sounds. To "bridge" our distance to this historical soundscape, the analytical framework sheds light on the historical relations between the material aspects of the soundscape (ecology of sound), interventions in sound (politics of sound) and the meanings involved (semiotics of sound).

Keywords: soundscape, sound, war, city, history, diaries

Introduction

> Nearly every war diary starts the same. In the early morning of 10 May 1940, the diarist wakes up to explosions, aeroplanes coming over or frightened relatives. He goes out onto the street in pyjamas, talks to neighbours, suggests the possibility of a military exercise, and then realises what is going on: the Netherlands is at war.[2]

That morning, it gradually dawned on the residents of Amsterdam that the resounding commotion no longer meant peacetime military exercises. Even if the sound of explosions or aeroplanes flying over may have seemed familiar, the realisation that this was not a military exercise made a world of difference. Undeniably,

---

1 This chapter draws heavily on the fifth chapter of my PhD thesis (in Dutch) *Het geluid van gisteren: Waarom Amsterdam vroeger ook niet stil was* (Maastricht University Press, 2014). I would like to thank Annemarie de Wildt, Hubert Berkhout, Jeroen Kemperman, Karin Bijsterveld and Andreas Fickers for their support during my research and their valuable comments on the thesis chapter. I am also very grateful to Ton Brouwers for his careful translation. Finally, I would like to thank the Dutch Foundation for Scientific Research (NWO) for funding the Soundscapes of the Urban Past project from which this article originated.
An adaptation of the thesis chapter has been published in a Dutch urban history journal ("Oorlogsgeluid in de stad en stadsgeluid in oorlogstijd: klank en betekenis", *Stadsgeschiedenis*, 7 (2012), 42–58). In addition, a brief adaptation in German has been published under the heading "Der Klang der Bezatzungszeit. Amsterdam 1940 bis 1945", in Gerhard Paul and Ralph Schock (eds), *Sound des Jahrhunderts*, 252–257.

2 Boom, *"We leven nog"*, 17.

the country was at war. The gunfire now took on a new meaning and was listened to differently. This dramatic shift can be compared, in some respects, with the experience of moving to another house: on the very first day, all the sounds and noises heard are new, and consequently much more noticeable. And that was exactly the situation in which the residents of Amsterdam found themselves on the morning the Second World War reached their city. It was as if overnight all the residents had moved to another house.

More than in Belgium or France, the invasion of the German army caused a shock in the Netherlands because during the First World War the country had remained neutral. Although scholarship on the Second World War has covered nearly all its aspects, the soundscape to which urban residents were exposed during those trying and dangerous years of war has largely been ignored. It consisted of hitherto unknown sounds, familiar sounds that took on different meanings, and sounds to which people attributed other symbolic values as the war evolved.

As a concept, "soundscape" refers both to our exterior world – the sounds that surround us – and to a construct designed to make sense of that world.[3] The constructive aspect implies that cultural meanings of sound and modes of listening are bound by place and time. As a result, historical recordings of specific sounds alone will not suffice when it comes to understanding how people perceived and judged particular soundscapes in the past. One also needs texts and other media containing information on people's views about and experiences of particular sounds.

Furthermore, the double meaning of the notion of the soundscape invites analysis from multiple vantage points. First, questions about sounds in particular environments are relevant; they pertain to the "ecology of sound". In addition, there are questions concerning the soundscape as a construct, which involve issues related to its meaning, or the "semiotics of sound". Finally, there are questions as to how sound is evaluated and functions in relationships of power. This aspect we may identify as the "politics of sound".[4]

There are two good reasons for studying the sounds of the city in wartime more closely. The first concerns the major role of sound in war experiences. The average war movie or computer game will bombard its viewers with sounds. Moreover, autobiographical records of life in the trenches during the First World War reveal how closely the experience of war is linked to particular sounds.[5] Thus far, little

3 Thompson, *The Soundscape of Modernity*, 1.

4 Jacobs, "The silence of Amsterdam".

5 One well-known book on life in the trenches is J. B. Priestley's *Margin Released.*

research has been conducted into sounds during wartime.[6] Carolyn Birdsall examines how sounds trigger war memories in her article "Earwitnessing: Sound Memories of the Nazi Period" (2009) and describes people's experiences during the air raids on Düsseldorf (more on this below) in her dissertation.[7]

The second reason for considering the soundscape of war in Amsterdam is the "collective neurasthenia" caused by the German occupation during the first three months, followed by years of fear and dread on one hand, and hope and an uncompromising stance on the other.[8] One way to gain access to the blankness or confusion caused by war is to consider the sensory experiences of everyday life. In many ways, war actually has a contradictory effect on a community. During the years of war, the people of Amsterdam became part of a much larger community experiencing turbulent worldwide events, symbolically reflected by the loud and recurring sounds of aeroplanes, air-raid sirens and anti-aircraft guns. At the same time, war can be highly disruptive of the everyday activities in a community. A city may grow increasingly still and empty, as if gradually returning to a simpler past, not as crowded or as noisy. Moreover, a focus on sensory experiences reveals the particular significance of sound in situations that are hard to judge or assess because of distance or darkness. If adults seem to be able to interpret the meaning of sounds quite rapidly in such situations, it is much harder for children, who tend to be troubled by loud or unknown sounds for a much longer time.

Yet how can we learn more about urban sounds in Amsterdam during the war? In the following section, I will specifically address war diaries – the main materials used in this chapter – as a valuable resource in this respect. Next, my argument will focus on the sounds that the residents of Amsterdam would hear during the day or night, and what the various sounds meant to them.

6 The work *Akustisches Gedächtnis*, edited by Robert Maier, focuses on war memories. Julia Encke's *Augenblicke der Gefahr* elaborates on the role of sound and its influence on the senses in life-threatening situations. Charles D. Ross's "Sight, sound, and tactics" addresses the role of sound in tactical warfare during the American Civil War. Jean Yaron writes about the transformation of the (symbolic) relationship between sound and the (assumed) power of armament in "'Silenced power'". Steve Goodman's *Sonic Warfare* draws attention to the impact of sound waves on the human body. Finally, the chapter on war sound in Greg Goodale's *Sonic Persuasion* addresses the use of sound in reports of war.

7 Birdsall, "Between noise and silence".

8 Meershoek, "Onder nationaal-socialistisch bewind".

Listening to war diaries

Which residents of Amsterdam were perceptive of sounds during the Second World War and left behind evidence of their experiences? Well, Dutch newspapers paid little or no attention to the local soundscape. When journalists wrote about the sounds of war, it was mostly about battles elsewhere, far away, or about issues of sound that were so neutral or "innocent" that they suited the media politics of the German occupier. After all, following the surrender of the Dutch army, newspapers and other media were immediately subjected to censorship.[9]

The best way of learning about the sounds of Amsterdam during the war is from diaries written at the time by local residents. The analysis in this chapter is based on a collection of thirty diaries from the archives of the Netherlands Institute for War, Holocaust and Genocide Studies (NIOD). Of course, keeping a diary is not a practice that equally appeals to people from all sorts of backgrounds, so my selection of thirty diarists does not reflect a cross-section of the Amsterdam population. Nevertheless, the authors in my sample represent various social classes, as well as various age categories. Moreover, nearly all the diarists seem to have shared the view of the overwhelming majority of people in Amsterdam that they would rather see the occupier leave the country sooner than later.

As a genre, the diary is hard to define.[10] Despite my decision to focus on war diaries evoking personal experience, my sample shows great variation in tone, style and level of detail. What all the diaries have in common, however, is that the author aims to report on life during the war in a more or less personal way and also writes down with some regularity what he or she finds relevant in that context. Although each author presents a unique perspective, they are all bound or influenced by common practices, views and linguistic conventions. This is potentially expressed in the selection of sounds that are written about and the words used to do so, and also in the way in which personal experience of sound is presented and judged. That of the thirty diaries, eight pay no attention at all to sound, sixteen occasionally, and only six on a regular basis supports the premise that sound is not an evident category for representing situations and experiences, regardless of the possible intensity of sounds.

9 Ibid., 241.

10 Cf. the following characterisation: "The diary, as an uncertain genre uneasily balanced between literary and historical writing, between spontaneity of reportage and the reflectiveness of the crafted text, between selfhood and events, between subjectivity and objectivity, between the private and the public, constantly disturbs attempts to summarize its characteristics within formalized boundaries". Langford and West, *Marginal Voices*, 8.

This interplay of historically determined practices, ideas and conventions makes it possible to compare individual or unique diaries in certain respects. As such they provide insight into the views and experiences of their authors, some of which they will have shared with their fellow residents. Careful analysis of the diaries, combined with historical information on events and activities that took place in the city, allows one to paint a picture of the Amsterdam "ecology of sound" during the war and how it evolved. Analysis of how the perception of sound was articulated during the course of the war offers insight into the meanings of particular sounds for local residents (semiotics of sound). Close analysis of passages that reflect their authors' judgments of individual sounds provides insight into the role of particular values or relationships of power (politics of sound).

Before the surrender

On the Friday before Whitsunday, in the night of 10 May 1940, the German army invaded the Netherlands. In the course of that night, many people throughout the country woke to the sound of aeroplanes and loud explosions. To most residents of Amsterdam, the war presented itself at first in the form of sounds:

> Woke up at around 3 o'clock to the roar of aeroplanes and explosions of anti-aircraft guns. Saw a dark-coloured aeroplane fly over that was shot at. Thought of an aerial attack on England, whereby our neutrality was violated. It fell silent, so we fell asleep again. At 4.30, woke up again and didn't sleep anymore. We were at war with Germany![11]

Over the following days, soldiers took up positions throughout the city, there were shootings in several places, and a stray bomb on the corner of Herengracht and Blauwburgwal caused dozens of fatalities and injuries. From her bedroom, Mirjam Levie, in her twenties, watched the bomb land and strike: "The screeching of that bomb was indescribable. Afterwards I was completely upset. I could no longer eat, looked as white as a sheet, and just sat there shaking all day".[12] The military attacks mainly targeted the port of Amsterdam and Schiphol Airport.[13] Until the surrender of the Dutch army, five days after the German invasion, to most residents of Amsterdam, the direct experience of the violence of war remained limited to sound, as is also clear from the diaries. Until 15 May, the noise of aeroplanes, gunfire and air-raid sirens is a constant in the diarists' accounts, serving in most cases as tangible evidence of their personal experience of acts of war.

---

11 Van Buuren, [diary], 10 May 1940.

12 Bolle, *Ik zal*, 19.

13 Meershoek, "Onder nationaal-socialistisch bewind", 239.

In this brief period, the diarists recorded the "roar", "drone" or "rumble" of aeroplanes; the "pounding", "booming" and "cracking" of artillery; and the "blaring" or "shrieking" of the air-raid sirens. These sounds were bothersome during the day, but especially at night: "Troubled sleep on account of the sirens; it has a fatiguing effect";[14] "We can't sleep because of the constant drone of the flying machines".[15] Oddly, perhaps, the diarists seemed most bothered by the air-raid sirens. Cannons were pounding or cracking and aeroplanes were droning or roaring, but only the sirens resounded with a "dismal blaring".[16] Moreover, the noise of aeroplanes and gunfire was not always explicitly linked to immediate danger, while the siren made people fearful or had a nerve-racking effect on them.[17]

> After six, the first siren went off. I thought our final hour had come, but I still went up the stairs quietly with my bundle across my arm, to make my toilet.[18]

For those in Amsterdam who were not directly involved in the warfare, those days were marked by a desire for information on the ongoing developments. On Friday, the war's first day, the stock exchange and schools were closed and all public entertainment was cancelled.[19] Next came Whitsunday weekend. Living between hope and fear, several diarists wrote about the sharp contrast between the fine spring weather and the anxiety about what was to come: "It was not warm, but still sunny outside. Nature rejoicing. Green, flowers opening, birds singing, but in all hearts oppression and fear".[20] Most diarists were constantly listening to the radio, and some went into the city centre to look for the latest news on the war. Rumours began to circulate, and every fifteen minutes there was radio news. Aside from the incidental sounds of aeroplanes, gunfire and sirens, it was mainly the radio that brought the residents of Amsterdam into contact with the war:

> In town, you could listen to the radio everywhere, most pleasantly in the hallway of the Cineac. Ongoing reports on aeroplanes and parachutists, who came down in all sorts of camouflages. [...] Several families decided to spend the whole night near the radio and not to undress.[21]

---

14 Van Buuren, [diary], 13 May 1940.
15 Bruijn-Barends, [diary], 13 May 1940.
16 Van der Does, [diary], 12 May 1940.
17 Bruijn-Barends, [diary], 13 May 1940.
18 Lans-Van der Wal, [diary], ± 12 May 1940.
19 Boer and Duparc, *Kroniek*.
20 Van der Does, [diary], 12 May 1940.
21 Lans-Van der Wal, [diary], 10 May 1940.

> At Freddy's, the radio was also on at Shabbat – as the chief rabbi had ordered – and you kept on hearing: "Post Wijk bij Duurstede (or some other place of course), twenty German planes identified, flying south-east to west, etc. etc." You felt the aeroplanes coming toward you. I went home again. An air-raid siren was now blaring away.[22]

On the afternoon of Tuesday 14 May, Rotterdam was heavily bombed. At the end of the afternoon, it was announced that the army leadership had decided to surrender. As one diarist wrote, "The situation is disastrous. 11 x siren. We can't hold out any longer. The queen gone, the government gone [...] At 7, we heard on the radio that we have surrendered. General Winkelman saved us from total destruction. We are deeply in trouble, and we are just staring blankly in silence".[23] Elsewhere in the city, the situation was quite tumultuous. Earlier that day, a flow of mainly Jewish refugees had tried to reach the port of IJmuiden – in vain: many did not even manage to leave town because of the clogged roads or reached IJmuiden too late and returned home.[24] Despite the beautiful weather, many chimneys were smoking, because papers and books were being burned in great haste, and also out on the streets. Many people had mixed feelings about the surrender.[25] "'The Netherlands has surrendered', we heard. The fear of new bombings like in Rotterdam was now over, but concerns about worse things arose".[26] In the early afternoon of 15 May, the residents of Amsterdam faced German troops marching through town. It was Hitler's wish to seal the German victory demonstratively with a parade through the city. The procession passed off without incident, and few diarists paid attention to it. Shmuel Hacohen, then aged fourteen, watched the invading army from the Berlagebrug: "We saw an awesome power, roaring past us endlessly".[27] From that time onward, the residents of Amsterdam would no longer merely *listen to* the war on the radio, in the form of air-raid sirens or as distant sounds from aeroplanes and gunfire.

After the surrender

After the capitulation, all sorts of local everyday activities were taken up in a different fashion, and this also had implications for the local soundscape. Businesses got going again on 16 May, and most schools reopened on Monday 20 May, but

---

22 Bolle, *Ik zal*, 20.
23 Bruijn-Barends, [diary], 14 May 1940.
24 Bolle, *Ik zal*, 21–22; Hacohen, *Zwijgende stenen*, 156.
25 Boom, *"We leven nog"*, 18–30.
26 Hanemann-Keleman, [diary], 1–6.
27 Hacohen, *Zwijgende stenen*, 157.

the sounds of traffic had changed. As most cars and busses were gone, there was a striking silence during the day: "May 18. [...] Amsterdam is very calm: there is no traffic and no damage and very few Germans about".[28] One week later, van Buuren writes: "Warm weather. Grocery shopping in the afternoon as usual. Virtually no cars anymore in the streets. Only the transportation of troops and groceries is still permitted. City bus services discontinued or reduced. Also doctors can no longer visit their patients by car".

From 9 July onward, as van Buuren reports, taxis returned to the streets. As fuel was scarce, private cars vanished permanently from the streets, while bicycle taxis were introduced. Horse-drawn carriages also served as taxis. After a visit to the Leidsche Plein Theatre, "to forget our worries", one author writes: "When we left [...] it was pitch black under the trees of the Leidsche Plein. With the silhouettes of people, horses and carriages (that now replace the taxis), the weak lights of arc lamps, bicycles and trams, the scene had an old-fashioned feel to it – last century or so".[29]

Horse-drawn carriages also replaced motorised freight traffic, in addition to handcarts and delivery bicycles. Because public transport was also heavily reduced when war broke out, we may presume that bicycle traffic increased considerably. More than two years after the capitulation, on 20 July 1942, van Buuren writes: "Amsterdam was a different city today than the day before yesterday. There were hardly any bicycles in the street, and it seemed odd".[30] The day before, a rumour had spread that in nearby cities the Germans had confiscated bicycles on a large scale, which caused many residents of Amsterdam to hide their bicycle in a safe place. Over half a year later, Sander writes: "It is increasingly striking how few people go to work in the city between 8 and 9 o'clock. But the trams are overcrowded, with countless people standing on the footboards [...]. The hundreds, or thousands, of bicycles are gone. If not confiscated by the German army, they are useless anyway, as there aren't any tyres".[31] A year and a half later, at the start of a winter of famine, the situation had changed once again: "The city is calm, with no trams; there is a great deal of cycling, and the thousands on reduced pay spend their leisure time cycling to or from Hoorn to try and find something to eat. This goes on despite the constant checks and confiscation of bicycles".[32]

It appears, then, that the city, already referred to as silent on 18 May 1940 by Cox, grew even more silent as the war progressed. This was caused not only by the

28 Cox, [diary], 18 May 1940.
29 Verwey, [diary], 5 September 1940.
30 Van Buuren, [diary], 20 July 1942.
31 Sander, [diary], 4 February 1944.
32 Van Donkelaar, [diary], 30 October 1944.

sharp reduction in motorised traffic and trams, on account of the mounting fuel shortage, but also because more and more people were arrested and deported. In late June 1942, Lans-Van der Wal writes: "I used to go out shopping for vegetables, but now I don't even bother, because it's all so bad; the persecution of the Jews is horrendous, but our boys all go as well. It's so quiet on the streets". Less than a year later, Mirjam Levie also concludes:

> On Munt square, it's so quiet that there are no police anymore. Cars are largely absent from Amsterdam, apart from those of the German army and a few others. There are also fewer bicycles, and all this has made traffic so sparse that traffic police have become absolutely superfluous. [...] But now the Jewish quarter, such as Breestraat, is deathly still, because many of the people have been transported away. [...] I wanted to tell you this because the city looks poor, deserted and dirty.[33]

The story of the increasingly silent city seems to contradict the fact that at the start of the war the number of "residents" suddenly increased. As one diarist observed: "The Germans marched in, singing their bloody clipped songs. The streets were closed down, the traffic halted, and the trams were overcrowded".[34] Three months after the capitulation, the city housed 14,000 German soldiers, quartered in schools and residential homes. Later, their number would decline.[35] One diarist commented that at night he could hear the German soldiers billeted in nearby schools.[36] None of the other diaries studied addresses the proximity of German soldiers or contact with them.[37] It seems as if they were neither heard nor seen, unless they shouted or were violent (more on this below).

According to the *Kroniek van Amsterdam*, over the period 1940–1943, the occupier and its Dutch collaborators (NSB [Nationaal-Socialistische Beweging], Jeugdstorm) organised several parades or marches in the city. In general, the diaries pay no attention to such events. Only in one very detailed diary do we find a comment to a newspaper clipping from 9 November 1940, headed "W.A.-March in Amsterdam". None of the diarist's acquaintances attended that march. This diarist also mentions a gathering held on 27 June 1941 at the Museumplein: "Today Seys Inquart is giving a major address here at the IJsclub. [...] Troops are marching

33 Bolle, *Ik zal*, 98.
34 Hanemann-Keleman, [diary], 1–6.
35 Meershoek, "Onder nationaal-socialistisch bewind", 241.
36 Van Buuren, [diary], 1 August 1940.
37 Other diaries in the NIOD collection contain passages on contact with German soldiers. In Boom's "*We leven nog*", there is a whole section on this, entitled "Allemaal nette kerels. De nieuwe machthebbers" [All of them decent guys. The new men in charge] (25–26). However, no reference is made to diarists from Amsterdam.

through the streets to attend. That the Germans attend we don't mind, but that Dutch traitors proudly march there singing miserably out of tune is awful and makes your heart freeze".[38]

The judgment on the singing of the German soldiers and NSB members in terms of "bloody clipped songs" and "singing miserably out of tune" is a reaction similar to not wanting to see or to communicate with the "enemy". It involves a typical example of the "politics of sound". The sound symbolises a situation that one abhors but cannot change, and so it is ignored or dismissed as awful and annoying.

Another example of the politics of sound is the emerging flow of anti-German songs and poems, as well as the measures taken by the occupier to combat it. *Het Parool* of 16 October 1940 ran a piece on half a dozen organ grinders from Amsterdam held in a German prison since early August "because they played national songs on their organ in the streets. *Das teutsche Vaterland kann ruhig sein!* At least these enemies are behind bars".[39] And on 6 November, one diarist, van der Does, wrote: "The flow of poetry continues – it is the only outlet we have".[40] Several diarists regularly quote such texts, and there are even diaries that consist nearly exclusively of lyrics, poems and jokes about the krauts. Formal holidays also served as occasions for protesting:

> Yesterday, I heard it was a mess in town on New Year's Eve. No permission was given to be out on the street after midnight, so people could not celebrate the N year with their friends. Still, at 1.40, our actual N. Year, there was great noise: loudspeakers from the windows, charivari, loud singing of the national anthem, etc. Shots were fired as well. They will not bring our people down.[41]

Conversely, on 6 July 1940, the Germans ordered the erection of a "bandstand on Dam square for concerts by the German Brass Band".[42] The "politics of sound" is expressed here in how the two camps dealt with music and national identity. Despite the great difference in power and means, both camps tried to avoid the music of the other and highly valued their own repertoire.

In the diaries, no further reference is made to street music and street organs. According to the *Kroniek van Amsterdam*, from 7 October 1942 onward, playing music in the streets was no longer allowed, and the use of street organs was also prohibited. After the noise of busy urban traffic, the Amsterdam soundscape

---

38 Bruijn-Barends, [diary], 9 November 1940 and 27 June 1941.

39 Pieter t. Hoen, "Het Deutschtum en de orgeldraaier", *Het Parool,* 1940.

40 Van der Does, [diary], 6 November 1940.

41 Bruijn-Barends, [diary], 4 January 1942.

42 Boer & Duparc, *Kroniek.*

lost another characteristic sound. That sound returned at the end of the winter, when residents were in desperate need of food: "Bands of children go from door to door and ask for food. In the shopping streets, begging women with children sing. People also go to the farms in the nearby polders. Dozens of street musicians appear in the streets".[43]

Likewise, another characteristic sound of the city, that of its bells and carillons,[44] was largely lost over the course of 1942 and 1943. On 5 January 1942, the start of the weekly carillon concerts on several towers was announced; six months later, the Germans decided to begin confiscating the carillons for the war industry.[45] On 11 July 1942, four days after going into hiding, Anne Frank writes in her diary: "Father, Mother and Margot still can't get used to the chiming of the Westertoren clock, which tells us the time every quarter of an hour. Not me, I liked it from the start; it sounds so reassuring, especially at night". Over a year later, the bells appeared to have been removed: "We've all been a little confused this past week, because our dearly beloved Westertoren bells have been carted out to be melted down for the war, so we have no idea of the exact time, either night or day".[46]

After the capitulation, there was a change not only in the sounds that could be heard, but also in the meaning or evaluation of certain sounds. In other words, a new "sound era" emerged. The anti-aircraft artillery, for example, targeted no longer the enemy but "friends", and so indicated that those friends were around. This changed the ways in which people listened to sounds, even if the sound itself or the potential risk of gunfire to urban residents did not change. As one diarist observed: "Wednesday 6 November 1940. Last night at eight the anti-aircraft artillery started pounding, also in the night at around one we were jolted awake by the guns. I always enjoy hearing it. What happens we don't know, nor will we find out for sure".[47]

Similarly, radio listening changed, when on 6 July 1940 the newspapers reported that it was "no longer permitted to listen to any foreign station except those under German control, in order to prevent the reception of incorrect reports".[48]

---

43 Herks, [diary], 8 March 1945.

44 "When in clear weather and bright sunshine one hears the city hall clock toll in the morning hours and the carillons sing from the city's bell towers, then it is hard to imagine the terrible dramas taking place in the sky and along the fronts" (Sander, [diary], 3 May 1943).

45 Boer & Duparc, *Kroniek.*

46 Frank, *The Diary of a Young Girl*, 26, 129.

47 Van der Does, [diary], 6 November 1940.

48 Van Buuren, [diary], 6 July 1940.

From this time on, people listened to foreign stations secretly, which was not without risks:

> I remember we were sitting on our heels in a room at our neighbour's, mister P.'s, leaving the room every now and then [...] walking around the house in the dark to see if anyone was near. [...] [We] were constantly looking up to check for anything unusual, our ears cocked, and a hand near the knob to turn off the radio at the first suspicious noise. There have been many cases of people caught for listening to the radio – five months imprisonment or worse, and high fines.[49]

Before and after the war, "neighbourly noise" referred to sounds from people living nearby that one did not want to hear – to noise *made by* neighbours. During the war, with one or two exceptions, the situation switched: what mattered now was that your neighbours did not hear any sounds *you made*. From Anne Frank's diary, it is clear that this vital condition determined life in the annex. Another dramatic example is found in the diary of Hanemann-Keleman, who took measures to have her baby daughter go into hiding when deportation seemed inevitable:

> Then I lift her from the box, whereupon my face quivers – immediately she starts crying. I force myself to smile because Bila's crying might be noticed by the neighbours, which must be avoided at all costs. Just a quick kiss on her sweet rosy cheeks, then Nies is off with the child.[50]

If neighbours still produced some particular sound, it was experienced as reassuring, rather than irritating, because "fortunately" it was only the neighbours. "Although we've calmed down, we're still startled by the merest sound. Last night, it seemed just like someone was coming up the stairs. Twice we left our bed to make sure it was nothing. Now, this afternoon, we were convinced for a moment that someone had opened our door. However, just like last night, it proved to be a noise from the neighbours".[51]

During the war, people's day/night rhythm was disrupted for several years in a row. After the capitulation, at the end of the first week of war, the air-raid siren remained silent until late August 1940, but the sound of dogfights or aeroplanes flying over could be heard during that period, day and night. On 20 May, van Buuren wrote that his wife was still listening to aeroplanes at night, but that he slept as usual again. In late May, his wife had her first good night's sleep again. In June, he occasionally heard shooting, at times heavy or nearby, but later a long way away again. He no longer noted down the sounds, and he wrote not

---

49 Verwey, [diary], 28 December 1940.

50 Hanemann-Keleman, [diary], 23 June 1943.

51 Koekenheim, [diary], 27 July 1942.

about aeroplanes, but about "activity in the air". Then, in the night of 6 July 1940, there was a fierce fight in the skies over Amsterdam: "Woke up at around 2.30 this morning to the heaviest anti-aircraft gunfire heard thus far. Shell fragments dropped into our street. At 4 o'clock, it started again just as fiercely".[52] At this time, as before the capitulation, van Buuren regularly referred to the sound of aeroplanes and gunfire. His choice of words underscores its repetitive nature: "At night, a constant drone of aeroplanes. Gunfire a long way away" (9 July); "A few air strikes with accompanying noise" (24 July); "Then the usual noise of aeroplanes droning and gunfire. After half an hour, a repeat of the whole episode" (27 July).

At times when the violence of war was experienced intensely, because it was new or close at hand, the diarists paid considerable attention to sound. Aeroplanes mostly "roared" and "droned", the siren "wailed" and the anti-aircraft gunfire "pounded". Together these sources of sound "boomed", particularly when they came so close that the sound could be felt or the house started quivering:

> This is the most awful night we've experienced yet. It's been so bad, so horrible that for the first time I was really scared, which doesn't happen easily. At half past 2, the first heavy bombs, directly followed by pounding fire, so heavy that you were shaking in bed. It seemed to last forever, and grew increasingly intense. Flashes of light danced up and down in front of our windows. It was far too dangerous to look out. Shell fragments dropped into the street; it was as if everything was taking place right above your head. In between the anti-aircraft gunfire, one could vaguely hear the aeroplanes droning. Whether there were any fights in the skies over the city, we don't know. You couldn't approach the windows because of the risk. Everything was booming and banging. It shook your nerves. It lasted until after 4, without interruption.[53]

With time, adult diarists seemed to get used to the sounds of war, even if they continued to be aware of the danger: "At 20.15, the gunfire aimed at the aeroplanes flying over started, and it went on, with brief interruptions, until 22.15. Two hours of booming and rumbling, but we hardly listen to it anymore. The radio plays on, and we read a book, without taking much notice of what goes on outside".[54] Air-raid sirens, aeroplanes and gunfire were now referred to with metaphors, such as "show", "concert" or "party". As one diarist put it, you hardly contributed to such a "party" by falling asleep.[55] Several diarists indicated that the sound of aeroplanes and sirens also prompted joy: "When there's an air-raid siren, you immediately

---

52 Van Buuren, [diary], 6 July 1940.
53 Bruijn-Barends, [diary], 6 July 1940.
54 Van Buuren, [diary], 27 January 1942.
55 Ibid., 27 March 1943.

hear cheering in the street: happily, they're back again!! No one's afraid. They won't get us, those krauts".[56]

Anne Frank, who was almost eleven years old when the war broke out, repeatedly indicated being afraid of what she mostly called "shooting". In the night, she would then look for comfort at her father's side (p. 89). On 2 June 1944, when she had been in hiding for nearly two years, she wrote of a "brand new prescription for gunfire jitters" (p. 306): when the shooting got loud, you should run up and down a wooden staircase, preferably stumbling every now and then, to take your mind off it. The noise you made would mask the shooting. Another diarist, Verwey, also indicates that his children were still very afraid of these sounds after he himself had grown used to them:

> Throughout the night, the wailing of the sirens and shooting – baby Ger crying all the time, Abje shaking in his bed, William with his eyes open – the whole family awake. [...] 8.30. Bombs falling, the siren wailing. The gunfire is pounding. We get used to it. Remarkably, we continue to work calmly, pretending that nothing is happening. Our sense of fatality has grown stronger.[57]

Aside from the sound of dogfights and bombardments, the Nazi government's terror gradually became associated with an array of frightening meanings attached to everyday sounds such as a doorbell or footsteps. One diarist comments that hearing "a footstep in the porch" triggers a general sense of fear: "When I hear something, it scares me to death. That's how it goes from 8 to 1 o'clock at night".[58] In her diary letter of 3 February 1943, Mirjam Levie concludes: "You don't know how completely we're thrown off balance. When the bell rings, we're startled. When there's a noise in the street, like a car banging, we wince. We pitied the German Jews when they were startled by every sound, but we've long been in the same situation".[59]

Not much has been written about the sounds accompanying round-ups. In most cases, of course, those who were carried off had no opportunity to write; even if they had, very few would have been able to write in such circumstances. Shouting soldiers and crying would have been part of it, but also silence and dumbfoundedness:

> Then, at 12, the loudspeakers arrive at the Merwedeplein. I prepare myself quickly. The doorbell rings: 2 policemen. First at Müller's, 120-er. Is allowed to stay. Manzi is also home and is allowed to stay. Next at our place. We're Lithuanians, and they don't know

56 Bruijn-Barends, [diary], 23 June 1943.
57 Verwey, [diary], 15 November 1940.
58 Spruit-Duis, [diary], 25 September 1942.
59 Bolle, *Ik zal*, 53.

> what to do with us. Come with us. Do you have any children? No. They go through the house, pull everything open, fortunately find nothing. […]
>
> I call to Müller and Manzi with a loud, hard voice: "Live well!". They don't respond. Fear – or shame about staying behind?
>
> The Merwedeplein is full of Jews and police. Long, sad lines of Jews carrying luggage had already walked across the Merwedeplein with children in pushchairs, holding their hands, sitting on their arms. Backpacks, bundles, blankets, headscarves; hats that go grotesquely with the backpacks. Crying, complaining and composed adults. And children! And the screaming German policemen, and the onlookers, more or less involved, and the NSB members.[60]

Those who stayed behind had difficulty dealing with it, lacking the knowledge we now have. That is why, from our perspective, texts from non-Jewish residents that show empathy with the Jews still seem painfully detached: "Tonight in our street, Jews arrested. A police van appeared, several Jews were put in, then it drove off again. No more planes flying over, it seems. Night after night passes and all is quiet!"[61] Also based on what we know now, however, such events are impossible to comprehend. In particular, Jewish residents of Amsterdam were confronted with absurd and degrading situations. One anecdote from a diary sums up the blatant absurdity of war: "Several months later, the parks were 'Out of bounds to Jews', and we got into trouble when our dog Tipsy was running and barking in Beatrixpark and refused to come back to us when we called him".[62]

Conclusion

On the level of the ecology of sound, the presented reconstruction of the Amsterdam soundscape shows two substantial changes in comparison to the pre-war situation. First, the various sounds – air-raid siren, aeroplanes, gunfire – constantly broke up the day/night rhythm during the war. This forced the residents, familiar with a pre-war lifestyle quite different from today's hectic round-the-clock lifestyle, to adjust and adapt to changing situations all the time: "It was again an old-fashioned bad night. We're never sure here, but after a few quiet nights you get a sense that 'now they'll finally leave us alone'. Now we've lost that confidence again".[63]

The abrupt or gradual disappearance of the familiar sounds of motorised traffic, people, street music, bells and carillons brought about a second significant change

60 Hanemann-Keleman, [diary], 23 June 1943.

61 Van Buuren, [diary], 1 October 1942.

62 Hanemann-Keleman, [diary], 1–6.

63 Bruijn-Barends, [diary], 30 October 1940.

in the soundscape. The noise of the city, its residents and its traffic triggered memories of bygone days: "There's such a scary silence in the streets, impending doom in the sky".[64] The city grew ever more still and seemed to be going back in time. As a result, the continuous heavy pounding of piling equipment, being heard all over Amsterdam during the reconstruction of the city after the war, invariably signalled a bright future. [65]

During the war, the politics of sound was mainly expressed in efforts from both the residents of Amsterdam and the occupier either to express their own national identity through music and songs or to ignore or ban the music of the opposite party. In the Amsterdam diaries, the sounds produced by the German occupier are ignored or despised as shouting and noise. A vivid example of judgments about the Germans being linked to the sound of their aeroplanes is the observation of the German machines' "puffing drone", while the British planes "are buzzing smoothly like blowflies".[66] Such descriptions correspond to how characters in naturalist novels are given an acoustic profile to indicate or judge their (social) position.[67] This again underscores the close link between the evaluation of a sound and of its source or cause.

As my analysis underscores, the specific articulations of the perception of sound provide insight into the semiotics of sound. The sound of shooting and gunfire is mentioned most in the diaries, followed by the sound of aeroplanes and air-raid sirens. In comparison to these sources, the diarists pay little attention to the sound of things like conversation, music, children playing, birds singing, radio, telephone or the weather. The words "calm" and "silence" occur regularly, but mostly to indicate that the noise of the gunfire, aeroplanes or siren has subsided. In most other cases, the mentioning of silence tends to have a negative connotation.

There are several explanations for the latter situation. First, the outbreak of war reduced the role of traffic noise or neighbourly noise, which, despite its nuisance, was rather associated with memories of better days. In the war years, silence also came to mean not knowing what was going to happen, as exemplified in the following quote: "The night was dreadfully silent: no flying machines, no anti-aircraft gunfire, no bullets, no cars on the roads, no trams or transporting of troops. You wake up realising that the fighting may have begun at sea, but the milkman doesn't

64 Ibid., 24 February 1941.
65 Bosscher, "De oude en de nieuwe stad". Amsterdam's soil is too soft to build on and requires piles to be driven deep into the ground as a foundation.
66 Lans-Van der Wal, [diary], 30 August 1940.
67 Schweighauser, *The Noises*, 72–73.

know and people simply go on with their business".[68] Finally, silence becomes a treacherous phenomenon if you do not want to be heard yourself.

The frequent mentioning of shooting, aeroplanes and sirens highlights the importance of those sounds for the residents of Amsterdam. Understandably so, because those sounds always signified potential danger, kept people from sleeping or interfered with their everyday activities. It is hardly surprising that during the same period, the soundscape of the German city of Düsseldorf was equally dominated by sirens, aeroplanes and bombardments.[69] Interestingly, however, a comparison of the two cities reveals that the evaluation of these typical war sounds reversed in both locations as the war progressed. In Amsterdam, the initial subdued reaction to these sounds was followed, once the danger came closer, by fear, and then by growing habituation, and even joy, because it perhaps signified imminent liberation. In Düsseldorf, the same sounds triggered euphoria in May 1940, but that later changed into fear and anxiety.[70] Secondly, the residents of both cities attributed another meaning and weight to the sound of air-raid sirens, aeroplanes, and gunfire or bombardments. In the diaries from Amsterdam, the sound of shooting and bombs was the most frequently mentioned sound. In Düsseldorf, the air-raid siren was mentioned most, it also being the sound that had the greatest impact on everyday life.[71] Conversely, residents of Amsterdam soon had doubts about the usefulness of the sirens, and they tended not to go into the shelters quickly.

> Five air-raid sirens! At ten, one, three, five and one other time, but I slept through it. The French anti-aircraft guns, which have only just arrived, are now operating. They're as good as the British, but they bark so terribly that it keeps people from sleeping. Most local residents will be puzzled as to why the sirens are wailing also in the middle of the night.[72]

The unreliability of the air-raid siren is also expressed in the slightly negative descriptions such as "wailing", "whining", "yelling" and "screeching". Likewise, the descriptions of gunfire targeted on "friendly forces" often come with a negative ring, such as "sputtering", "blaring awfully", "droning" and "barking". The residents of Amsterdam liked to hear aeroplane "fly-overs", especially if they kept a distance. When they approached, they were described as "roaring" and "droning", or, further away, "whirring", "humming" and "buzzing". Yet the words for the sounds of war were not highly negative, which justifies the conclusion that adults managed

68 Verwey, [diary], 9 September 1940.
69 Birdsall, "Between noise and silence", 150.
70 Ibid., 144.
71 Ibid., 144–145.
72 Van der Does, [diary], 8 November 1940.

to adapt quickly to the changing soundscape. Considering the experiences of the Verwey children and of Anne Frank, however, one should not underestimate what it means to children and youngsters to "move to a new soundscape".

Bibliography

Diaries

All sources from the Instituut voor oorlogs-, holocaust- en genocidestudies (NIOD, Institute for War, Holocaust and Genocide Studies), unless otherwise stated.

Anonymous, NIOD no. 1419.

Anonymous, Wim, NIOD no. 1164.

van den Berg, NIOD no. 1264.

Beuth, NIOD no. 1134.

Blitz, NIOD no. 1242.

Bolle, Mirjam, *Ik zal je beschrijven hoe een dag er hier uitziet: dagboekbrieven uit Amsterdam, Westerbork en Bergen-Belsen* (Amsterdam: Contact, 2003).

Bruijn – Barends, NIOD no. 1151.

van Buuren, NIOD no. 1230.

Cox, NIOD no. 1082.

van der Does, NIOD no. 1179.

van Donkelaar, NIOD no. 1570.

Frank, Anne, *The Diary of a Young Girl*, tr. Susan Massotty, ed. Otto H. Frank and Mirjam Pressler (Amsterdam: Bakker, 2008).

Gerlagh, NIOD no. 1539.

Goudsmit, NIOD no. 1205.

Hanemann-Keleman, NIOD no. 1133.

Hemsing, NIOD no. 1225.

Herks, NIOD no. 1129.

Hirschel, NIOD no. 1433.

Koekenheim, NIOD no. 1208.

Koekenheim-Voet, NIOD no. 1208.

Koopman, NIOD no. 1160.

Korting, NIOD no. 1456.

Lans-Van der Wal, NIOD no. 1092.

Luthart, NIOD no. 1313.

Meents, NIOD no. 1136.

Polak, NIOD no. 1131.

Roos-van Rossem, NIOD no. 1513.

Sander, NIOD no. 1141.

Spruit-Duis, NIOD no. 1186.

Verwey, NIOD no. 1187.

Other literature

Birdsall, Carolyn, "Earwitnessing: Sound memories of the Nazi period", in Karin Bijsterveld and José van Dijck (eds), *Sound Souvenirs: Audio Technologies, Memory and Cultural Practices* (Amsterdam: Amsterdam University Press, 2009), 169–181.

– "Between noise and silence: Sound, technology and urban space during Nazi Germany", PhD diss., University of Amsterdam, 2010.

Boer, J. F. M. de, and Duparc, S., *Kroniek van Amsterdam over de jaren 1940–1945* (Amsterdam: Genootschap Amstelodamum, 1948).

Boom, Bart van der, *"We leven nog": de stemming in bezet Nederland* (Amsterdam: Boom, 2003).

Bosscher, Doeko, "De oude en de nieuwe stad", in Piet de Rooy et al. (eds), *Geschiedenis van Amsterdam: tweestrijd om de hoofdstad 1900–2000* (Amsterdam: SUN, 2007), 337–397.

Damousi, Joy, and Deacon, Desley (eds), *Talking and Listening in the Age of Modernity: Essays on the History of Sound* (Canberra: ANU E Press, 2007).

Encke, Julia, *Augenblicke der Gefahr: Der Krieg und die Sinne* (Munich: Wilhelm Fink, 2006).

Goodale, Greg, *Sonic Persuasion: Reading Sound in the Recorded Age* (University of Illinois Press, 2011).

Goodman, Steve, *Sonic Warfare* (Cambridge MA: MIT Press, 2010).

Hacohen, Shmuel, *Zwijgende stenen: herinnering aan een vermoorde jeugd* ('s-Gravenhage: BZZTôH, 1990).

Hoen, Pieter t., "Het Deutschtum en de orgeldraaier", *Het Parool*, 16 October 1940.

Jacobs, Annelies, "Analyzing the dramatization of sound in the urban past", paper presented at the Blankensee Colloquium, Berlin, 17–19 June 2010.

– "The silence of Amsterdam before and during World War II: Ecology, semiotics, and politics of urban sound", in Daniel Morat (ed.), *Sounds of Modern History: Auditory Cultures in 19th- and 20th- Century Europe* (New York: Berghahn, 2014), 305–321.

Langford, Rachael, and West, Russell, *Marginal Voices, Marginal Forms: Diaries in European Literature and History* (Amsterdam: Rodopi, 1999).

Maier, Robert (ed.), *Akustisches Gedächtnis und Zweiter Weltkrieg* (Göttingen: V&R unipress, 2011).

Meershoek, Guus, "Onder nationaal-socialistisch bewind", in Piet de Rooy, et al. (eds), *Geschiedenis van Amsterdam: tweestrijd om de hoofdstad 1900–2000* (Amsterdam: SUN, 2007), 235–321.

Paul, Gerhard, and Schock, Ralph (eds), *Sound des Jahrhunderts: Geräusche, Töne, Stimme 1889 bis heute* (Bonn: Bundeszentrale für politische Bildung, 2013).

Priestley, J. B., *Margin Released: A Writer's Reminiscences and Reflections* (London: Heinemann, 1962).

Ross, Charles D., "Sight, sound, and tactics in the American civil war", in Mark M. Smith (ed.), *Hearing History: A Reader* (Athens, GA: University of Georgia Press, 2004), 267–278.

Schweighauser, Philipp, *The Noises of American Literature, 1890–1985: Toward a History of Literary Acoustics* (Gainesville: University Press of Florida, 2006).

Smith, Mark M. (ed.), *Hearing History: A Reader* (Athens, GA: University of Georgia Press, 2004).

Thompson, Emily, *The Soundscape of Modernity: Architectural Acoustics and the Culture of Listening in America, 1900–1933* (Cambridge, MA: MIT Press, 2002).

Yaron, Jean, "'Silenced power': Warfare technology and the changing role of sounds in twentieth-century Europe", *Zeithistorische Forschungen/Studies in Contemporary History*, 2017/2, http://www.zeithistorische-forschungen.de/2-2011/id=4541, accessed 12 July 2017.

Uta C. Schmidt

University of Duisburg-Essen (UDE), Netzwerk Frauen- und Geschlechterforschung NRW

Roaring war and silent peace? Initial reflections on the soundscape in the Ruhr between area bombing and reconstruction

Abstract: In this chapter, I explore how the Ruhr area of North Rhine-Westphalia (Germany) sounded around 1945, from the battle sounds of the Allied carpet bombing, through the atmospheres of vanished cities and the soundscape of bunker life, to the hot jazz bands that symbolised a new era of humanity, tolerance, freedom and democracy after the Nazi dictatorship. I thematise the traumas endured by people at that time, who woke up every night screaming, with the wartime experiences still hammering in their heads, and recall the sounds of reconstruction. In relating the dreadful sounds of the Allied squadrons, West German citizens emphasised the stunning evidence of crimes committed not by the Germans against others, but by others against Germany. The thesis presented here is that their auditory experiences of night-time bombing in the West nourished an identity as victims – here as victims of the Allied bombing campaign. This is a corresponding story, which derives its own persuasive power particularly from the remembered sounds of the reconstructed industries, which indicated: "We're back!" It becomes clear that an orientation towards a fixed polarisation between "loud war" and "quiet peace" does not work. In this chapter, I argue for simultaneous thinking about the non-simultaneous, for overlapping soundscapes and for case studies of examples, situations, details and fragments, aware of anthropological and technical, as well as spatial, references.

Keywords: Ruhr, carpet bombing, reconstruction, hot jazz, epistemological question

On 1 September 1939, Nazi Germany invaded Poland. This aggression resulted in the outbreak of the Second World War. On 10 September 1940, Nazi Germany invaded France and Belgium. Dorothea Günther, hiking with her friend on the day of the western raid, remembered:

> This year, there were an incredible number of cockchafers. Frozen by the cold night, they lay on the road. We couldn't avoid them at all. When we stepped on the beetles, their bursting chitinous exoskeleton made a terrible noise. For me, the beginning of the Battle

of France was always associated with this noise, even more so when the war reports on the radio euphorically spoke of "cracking the tanks".[1]

German has the onomatopoeic word "knacken", and you "knack", or crack, tanks. Dorothea Günther remembered the beginning of the Western campaign through a dense acoustic metaphor. In her short account, she addressed five acoustic dimensions. They all touch in their own way on the epistemic challenges of historical sound studies:

1) There is the acoustic event of the bursting chitinous outer wings, where sound is a question of physics or deformation technology.
2) In extreme situations, when human perception is particularly acute, sound is a question of cognition.
3) For literary studies, the acoustic experience becomes a metaphor in the narrative of the Western campaign.
4) Media studies should be interested in the correspondence between individual acoustic experiences and amplification by media-transmitted sounds, such as the onomatopoeic "Panzer knacken" ("to crack tanks"), which serve as mnemonic triggers to initiate meanings associated with what has happened. Sounds that we detect in a particular situation are anchored. We link them to meanings in our lives. Sounds can preserve such moments.
5) In the cultural memory of the Germans, "cockchafers" and "war" are profoundly linked. The children's song "Maikäfer flieg" [Cockchafer fly], handed down since the early modern era, articulates childish fears of loneliness. It transmits the horrors of war to subsequent generations.[2]

---

1 "Es war ein an Maikäfern sehr reiches Jahr; durch die Nachtkälte erstarrt lagen sie auf dem Weg. Nicht allen konnten wir ausweichen, wir traten hin und wieder darauf und der berstende Chitinpanzer verursachte ein entsetzliches Geräusch. Für mich blieb der Beginn des Westfeldzuges immer mit diesem Geräusch verbunden, um so mehr, wenn in den Kriegsberichten von Panzer knacken die Rede war". Dorothea Günther, in http://www.dhm.de/lemo/zeitzeugen/dorothea-günther-der-krieg-193940, accessed 29 March 2017.

2 "Maikäfer flieg, Vater ist im Krieg, Mutter ist in Pommernland, Pommernland ist abgebrannt, Maikäfer flieg"; "May bug fly, father's at war, mother's in Pomerania; Pomerania's burned, may bug fly". See Wieden, "Maikäfer, flieg!". The melody of the song corresponds to that of the lullaby "Schlaf Kindlein, schlaf" [Sleep my child, sleep]. In 1974, Anselm Kiefer, an internationally renowned artist, produced a painting titled *Maikäfer flieg!* [May Bug, fly!], oil on burlap, 220 cm x 300 cm. In 1991, he reprised the topic and the title in the lead work *Maikäfer flieg!* [May Bug, fly!], emulsion, ash and

So historical sound studies concern both the objective and the subjective dimension of sound, its physical and cultural amplitudes. They address issues relating to the French historian Roger Chartier's concept of "representation".[3] They prompt numerous questions. How can we reconstruct the relationship between immediacy and mediacy? How can we describe acoustic "atmospheres" as the relations between our surroundings and our existential outlook? They benefit from a threefold focus: the linguistic, the spatial and the material.

To be clear, this article by no means advocates giving up all the visual in favour of regaining the acoustic. It follows the German philosopher Gernot Böhme's argument that we should study urban history through more senses and attach greater value to them.[4] Michael Bull and Les Back argued: "Thinking with our ears – a translation of Theodor W. Adorno's phrase 'mit den Ohren denken' – offers an opportunity to augment our critical imagination, to comprehend our world and our encounters with it according to multiple registers of feeling".[5]

How did the Ruhr sound around 1945? This question, concerning the sounds during the period of area bombing, liberation and post-war Germany, is posed here for the first time. Approaching the issue from the perspective of the history of everyday life and the regional history of the Ruhr area, this article focuses on the aural phenomena inherent to and imposed on this region, especially on Dortmund.

The Ruhr is an agglomeration of 4,434 km^2 in North Rhine-Westphalia. It is a complex of industries and transport systems built during the industrial era, with high population density in urban areas. Its specific sound space was defined by heavy industries, traffic and daily life. Recent urban studies describe the Ruhr area as an outstanding conurbation. It was well known as the "Armory of the German Reich". Its final elimination by Allied air strikes led to the end of the war in Europe.

My approach to its soundscape is epistemologically orientated towards interpretative cultural anthropology, which has long benefited from Clifford Geertz's "thick description".[6] The French historian Alain Corbin defined sound history – or the history of the aural – as a story of social conflicts.[7] Before that, however, Murray Schafer had sensitised historians to the social functions of sounds. Even though he

---

lead on canvas, 330 x 560 cm; see exhibition catalogue, Schneider and Kiefer, *Anselm Kiefer*, 88. I am grateful for this information to Renata Tańczuk of Wrocław.

3 Chartier, "Kulturgeschichte"; Chartier, "Die unvollendete Vergangenheit".

4 Böhme, *Anmutungen*; Böhme, *Atmosphäre*; see also Feiereisen and Hill (eds), *Germany in the Loud Twentieth Century*, 6.

5 Bull and Back (eds), *The Auditory Culture Reader*, 2.

6 Geertz, "Thick description".

7 Corbin, *Les cloches*.

is now critically debated,[8] his book *The Tuning of the World*, published in German in 1988 as *Klang und Krach*, paved the way for sound studies in Germany.[9]

Roaring war

After the armistice with France was signed, on 22 June 1940, all the bells rang for one week at lunchtime, for a quarter of an hour. The political system instrumentalised church bells and their symbolic capital for stirring the country into a national frenzy.

There is one thing we must be clear about here: only a few hours after the German raid on France of 10 May 1940, British and US-American bombers began air attacks on the region between the Rhine and the Ruhr. So their air strikes pre-dated the German "Blitzkrieg" and the Battle of Britain.

Figure 1: Leaflet dropped over Essen after a British air raid in May 1943, entitled "Fortress Europe has no roof", Imperial War Museum, London

Die Festung Europa hat kein Dach

IM April 1943 warf die R.A.F. mehr als 10 Millionen Kilogramm Bomben auf deutsche Industrieziele.

Im Mai 1943 wurden über 12 Millionen Kilogramm Bomben abgeworfen.

In einer einzigen Woche im Mai fielen 750 000 Kilogramm Bomben allein auf das Ruhrgebiet.

Bis 1. Juni 1943 haben englische Flugzeuge über 100 Millionen Kilogramm Bomben auf Deutschland abgeworfen.

Die Royal Air Force ist heute stärker als die deutsche und italienische Luftwaffe zusammen.

In Amerika wurden allein im April 1943 7 000 Flugzeuge fertiggestellt.

Die amerikanische Flugzeugindustrie hat am 31. Mai das 100 000 Flugzeug für diesen Krieg geliefert.

Amerika produziert heute mehr Flugzeuge als Deutchland, Italien und Japan zusammen.

Das sind die Tatsachen. Was folgt daraus? Man könnte sagen: Was die Engländer ausgehalten haben, können die Deutschen auch aushaltern.

Aber der Vergleich hinkt. Die englisch-amerikanische Luftoffensive gegen Deutschland ist bereits jetzt weit schwerer als die deutsche Luftoffensive gegen England je war. In ihrem schwersten Angriff auf eine englische Stadt (London, 10. Mai 1941) warf die Luftwaffe 450 000 Kilogramm Bomben in einer Nacht.

In ihrem bisher schwersten Angriff auf eine deutsche Stadt (Dortmund, 23. Mai 1943) warf die R.A.F. 2 000 000 Kilogramm Bomben in einer Nacht. Mehr als viermal soviel — und das ist erst der Anfang.

Die Engländer konnten standhalten, weil sie wussten: Wenn sie standhalten, musste es besser werden. Sie hatten kaum angefangen zu rüsten. Sie kamen erst in Gang. Sie wussten, dass ihre eigene Luftmacht und Luftverteidigung von Tag zu Tag wuchs.

Sie sahen amerikanisches Kriegsmaterial in Massen ankommen und sie konnten hoffen, dass Amerika eines Tages als Verbündeter an ihrer Seite stehen würde.

Was haben heute die Deutschen zu hoffen, wenn sie sich standhaft weiter bomben lassen?

Neue Verbündete gibt es nicht mehr. Deutschlands eigene Kampfkraft lässt nach. Die englisch-amerikanische wächst von Tag zu Tag. Deutschlands Fabriken werden bombardiert. Seine Produktion ist um 20% gesunken. Amerikas Produktion ist „bombensicher".

Deutschlands geschwächte Luftwaffe muss an drei Fronten kämpfen. Überall ist sie in der Defensive. Die R.A.F. und die amerikanische Heeresfliegerei haben, trotz Atlantik-Wall, eine Front im Herzen Deutschlands geschaffen. Die „Festung Europa" hat kein Dach.

Daran ist nichts mehr zu ändern. Die deutschen Arbeiter und Ingenieure, die in den Fabriken aushalten, können den Gang des Krieges nicht mehr wenden. Sie können den Krieg nur verlängern. Das heisst: Sie können dafür sorgen, dass noch mehr Bomben auf Deutschland fallen.

Es werden immer mehr werden von Monat zu Monat.

Wie lange soll das noch so weiter gehen ?

8 Morat, "Sound Studies–Sound Histories", 5; Missfelder, "Period Ear", 37.
9 Schafer, *The Tuning of the World*.

People in the Ruhr began to live with an air raid alarm system, which indicated different levels of threat: "Public air warning", "Pre-warning", "Air raid alarm", "Before All Clear" and "All-Clear". The system became so complex that it had to be continuously explained to the population through the mass media. The signals were triggered increasingly often until the alarm system imploded: companies categorically protested against the proliferation of the air raid alerts, as they disrupted production. During the war, the alarm system was adapted over and over. From September 1944 on, the signal "Acute air threat" was introduced to indicate the highest level alert strictly and unambiguously.[10]

"Suddenly, as in the morning, the sirens indicated 'Acute air threat', again without pre-warning".[11] People learned to find their bearings in space and time thanks to the signals: continuous tones, increasing and decreasing, high-low, sinusoidal oscillation between pitches of about 200 Herz and 400 Herz, interrupted by pauses.[12]

From mid October 1944, both the British Bomber Command and the USAAF 8th Air Force revisited the Ruhr with squadrons of well over 1,000 aircrafts. Towns and cities that had already been substantially damaged again suffered carpet bombing.

---

10 Blank, "Kriegsendphase", 413.

11 A recollection of Helga Pflugstedt, in Grontzki et al. (eds), *Feuersturm und Hungerwinter*, 88.

12 See Dietrich Jansen, http://bunkermuseum.de/akute-luftgefahr-fuer-emden-und-meldungen-des-drahtfunks/, accessed 29 March 2015.

Figure 2: Map of grid squares for air raid warnings, source: Bunkermuseum Emden

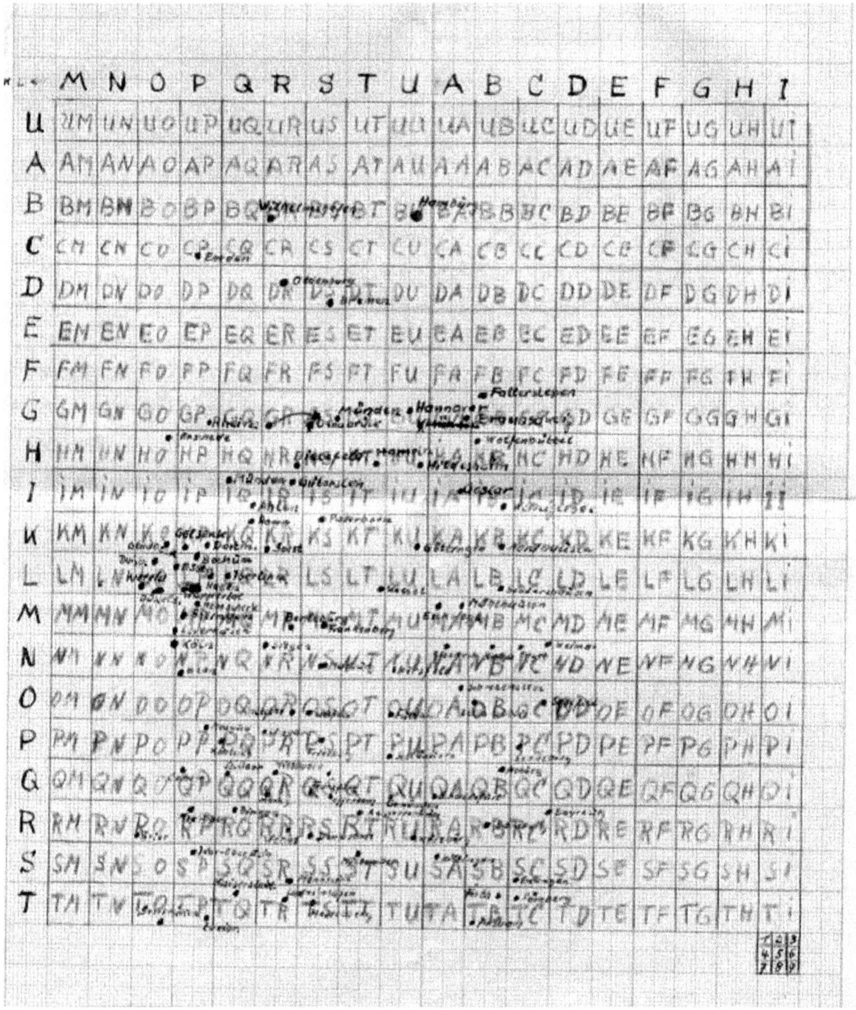

The war became condensed to the acoustic cipher "Konrad Paula 8": "Konrad Paula 8 – those coordinates became the key sound for the flight path of the Royal Air Force to Herne, announced by radio", Christel S. told me.[13] All the participants

13 Schmidt, "Radioaneignung", 350

in present-day witness projects talk about this sounding cipher given by radio – interestingly, without being asked about it.[14] They also recall the sound of the message: "Wir geben eine Luftlagemeldung" ("We're giving an air position report"), or "Achtung, Achtung! Fliegerverbände nähern sich dem Ruhrgebiet!" ("Attention attention! Air formations at the approach to the Ruhr").

Children creatively turned the messages into counting-out rhymes: "Achtung, Achtung, Ende, Ende/über'm Kuhstall stehn Verbände/über'm Schweinestall die Jäger/Achtung, Achtung, Schornsteinfeger" ("Attention, attention, Roger, Roger/ in the sky over the cowshed there are fighter squadrons/in the sky over the pigsty there are interception planes/Attention, attention, chimney sweep").[15] People learned to listen closely, assessing the risks and listening to the sounds of the German anti-aircraft guns (Flaks), distinguished from the battle sounds of the Allies. They learned to decode all sounds to differentiate between safety and extreme danger and the hostile situation of the bombardments.

Joseph P. Krause was twelve years old at the time, living in Schalke, a district of central Gelsenkirchen. At 13.47, the broadcast announced:

> "Attention! Attention! Strong enemy bomber formations at the approach to Gelsenkirchen". Even as this message was being broadcast, all hell broke loose. We fled from the carpet bombing, with the public sirens of "acute air threat" heard in the open […]. Through the hail of bombs and between flying debris and flak shrapnel, we made it to the public air raid shelter […]. Tens of thousands of explosive and incendiary bombs rained down on Schalke. All supply lines were immediately interrupted. No water. No power. No radio signals or warnings. […] The bombs were fitted with rattling and whistling air screws to increase the horror effect among the civilian population. Thanks to the perfidious acoustics of the bombs, we sensed in advance when a bomb would hit our neighborhood and how heavy it would be, and we instinctively ducked and crouched on the floor […]. Animal screams of agony came from everywhere. Children and women were crying hysterically, cursing and praying loudly, throwing themselves on the ground, whimpering, pleading in vain for mercy to the invisible God. We were prisoners in hell. Smoke. Heat. Then an infernal roar and crackle. The building on us had collapsed.
>
> This was Dante's "Inferno" pure and simple. There was weeping and gnashing of teeth. The battered people roared and sobbed in despair and fear of death. […] Even today,

14 In the mid 1990s, I explored the field of oral history to gain insight into broadcasting, everyday life and war. See Schmidt, "Radioaneignung", 243–360. In 2013, with my colleague Susanne Abeck, I began researching historical consciousness in the Ruhr. We are collecting experiences, memories and personal stories. On tour with our book *Stulle mit Margarine und Zucker* [Bread with margarine and sugar], people come to us, telling their stories. Most of them are directly related to wartime childhood.

15 Recollection of Jürgen Weber, in Abeck and Schmidt (eds), *Stulle*, 61.

> after more than 60 years, I weep like an old man when the calendar indicates the 6th of November. We were driven by burning, unconstrained horror and terror. We had survived the eruption of hell, volcanoes of phosphorus, attacks of fire accelerants, the mighty blasts of air mines, the roar of a thousand fires.[16]

Karlheinz Urmersbach, born in 1928, remembered 6 November 1944 in Gelsenkirchen:

> The four-engine British bombers arrived. It was horrible: the hum of the squadrons, flying very low one behind the other. The hum of the engines set our teeth right on edge. We knew from experience that the sinister roaring and whistling would start immediately. But before we ran into the shelter, we were surprised about our Flak, which didn't fire with the 10.5 and 8.8 cm shrapnel […]. Then the bombing began, with rattling and crashing […]. Everyone began to pray in fear […].[17]

In all bunker stories from the Ruhr, the first thing one notices is the praying, remembered as a prominent part of the soundscape. Here we can find spontaneous, not ritualised, religious patterns in social practices, not found in other circumstances.

> Above all, I keep in my ears the terrifying noise that caused the bombing, although I was still such a small girl. Yelling, whistling and rumbling, the dull bombings that shook the thick walls, mingled with the cries of children and the quiet weeping and praying of the adults.[18]

> Women and children were screaming, prayers were said.[19]

> We felt the violent impact like a rolling, thunderous wave. Those were the bombs […]. People wept and prayed to God […].[20]

Sometimes the memories have a more literary-style narrative: "Squatting and crouching, they sat on the uncomfortable wooden benches listening to the sound that penetrated the thick walls from outside. Were there bombs falling anywhere? Are the strikes far away or nearby? Here you have to learn praying again! Aunt Martha, however, didn't go down into the air raid shelter. She had an unshakeable faith in God. 'Nothing will happen, the Blessed Virgin Mary will stand by me', she said. Oh, yesterday she dedicated a candlelight to her, didn't she?"[21]

---

16 Joseph P. Krause, http://ww2today.com/6-november-1944-black-monday-for-gelsenkirchen-as-raf-bombers-return, accessed 29 March 2015.

17 Sixty-five years after the bombing of Gelsenkirchen, Karlheinz Urmersbach (b. 1928) spoke of his experiences on 6 November 1944, http://www.gelsenzentrum.de/bombenangriff_1944.htm, accessed 29 March 2017.

18 Gerda Pedack, in Abeck and Schmidt (eds), *Stulle*, 69.

19 Herbert Pemp, in ibid., 70.

20 Paul Kruck, in ibid., 73.

21 Brigitte Wiers, in ibid., 71.

Figure 3: Kampstraße in Dortmund, with the destroyed Petrikirche, in the background the Union Brewery, photo: Dortmund municipal archives

The remaining population, including many thousands of slave labourers and prisoners of war, suffered the final throes of the Second World War as carpet bombing. However, "The danger came not only from the squadrons, but also from single low-flying aircrafts with their guns shooting", recalled Christel S.[22]

The Italian military chaplain Guiseppe Barbero lived in the POW (Prisoner-of-war) camp Stalag VI D. No shelters were provided for prisoners of war and

22 Interview with Christel S., at Herne, 10 September 2016.

forced labourers. They were not allowed to use the air raid shelters, which were reserved for the indigenous population. On 12 March 1945, Barbero barely escaped with his life: "The hellish four-engine bombers come close [...]. We were no longer in a hailstorm of cluster bombs, but of bombs of an unheard-of calibre. Bombs right, bombs left, 5 or 10 meters away from us, the hissing bombs fell in thick clusters and exploded with an unconceivable roar that we had never heard before [...]."[23] On 13 April 1945, at 4:30 p.m., the Second World War ended in Dortmund. The Dortmund War Chronicle reported: "The day of the occupation and the bombing has stopped".[24]

Figure 4: The city of Dortmund, 10 May 1945, photo: German Federal Archives

To help imagine the Dortmund soundscape on 12 March 1945, consider the following numbers: more than 1,000 mostly four-engine machines, Lancasters, Halifaxes and Mosquitos, dropped 5,000 tons of mines and explosive bombs. The

23 Giuseppe Barbero, in Blank, "Kriegsende", 9.
24 Luise von Winterfeld, Kriegschronik 1939–1945, Stadtarchiv Dortmund lfd. Nr. 169, "Dortmunder Chronik des Jahres 1945", *Heimat Dortmund*, 1 (2015), 65.

attack lasted forty minutes. First the residents could hear the German Air defence command ("Flak"), which was stationed at Dortmund-Wellinghofen. But after fifteen minutes, the Flak ceased.[25]

Interpreting the aerial view of the destroyed city taken by the Allies as a source for urban sounds, many questions arise. How does an area of rubble, a landscape of ruins sound? How does a landscape sound when there are no longer any rattling trams, rows of houses, pulsating traffic, children's voices in schoolyards, pavements echoing to hasty steps, bell towers showing the time, factory sirens, steel hammers and cargo transports? How to reconstruct the acoustic relations between environment and existential outlook when the atmosphere of a city has vanished, when private and public spaces no longer exist? These issues concern theory as well as practice: museums and history projects increasingly ask for sound installations that convey this atmosphere of ruins to generations with no wartime experiences.

Master narratives

People recalling their childhood memories at the end of their life remember the sounds of warning like "Konrad Paula 8", "acute air threat" signals, squadrons, rattling and whistling air screws, bombs, crying and praying, but they do not remember any sounds of the new city of Dortmund. The soundscape of the city undergoing reconstruction changed imperceptibly. This corresponds to Murray Schafer's categorisation of keynote sounds, signals, soundmarks and archetypal sounds,[26] and it also tallies with the thesis that sounds only catch the ear when they change, arise or disappear radically.[27]

However, they did remember the optimistic sounds of restarted coal, steel and iron production, familiar sounds that indicate a return to industrial power, symbolised by factory sirens, mine shaft signals and the rhythm of steel hammers. In the collective memory of the Ruhr, these sounds indicated: "We're on the way back!"

---

25 Blank, "Kriegsende", 8.

26 Schafer, *The Tuning of the World*, 9–10.

27 Missfelder, "Period Ear", 37.

Figure 5: Pits and plants as symbols of the economic miracle, here pit "Minister Stein" in Dortmund-Ewing, photo: Kultur- und Geschichtsverein Minister Stein

The bunker population was united by the fear of death, which was expressed by metaphors of sounds in the testimonies. In telling bunker stories of bombing, crying and praying, West Germans emphasised "the stunning evidence of crimes committed not by the Germans against others, but by others against Germany",[28] as Robert G. Moeller recalled when speaking of the end of the war in the East. The sound experiences of the bombing nights in the West also created an "imagined community". They nourished an identity as victims – here as victims of the Allied bombing campaign. This is a corresponding story, which derives its own persuasive power particularly from the sounds of the reconstructed industries.

During the reconstruction phase after the Second World War, the Ruhr was the motor of the western German economic revival. The mining, iron and steel industries helped the country to repair the wartime damage. The significance of the soundscape of reconstruction, linked with the soundscape of the victims, brilliantly realises the German national master narrative of the "Wirtschaftswunder" – the

28 Moeller, "Remembering the War", 84.

economic miracle. In fact, the rhythm of the endless Ruhr coal wagons became the soundtrack of the economic miracle, a master narrative determining the identity of the Ruhr up to the present day.

The new Dortmund

Figure 6: Old and new city layout, source: Dortmund municipal archives

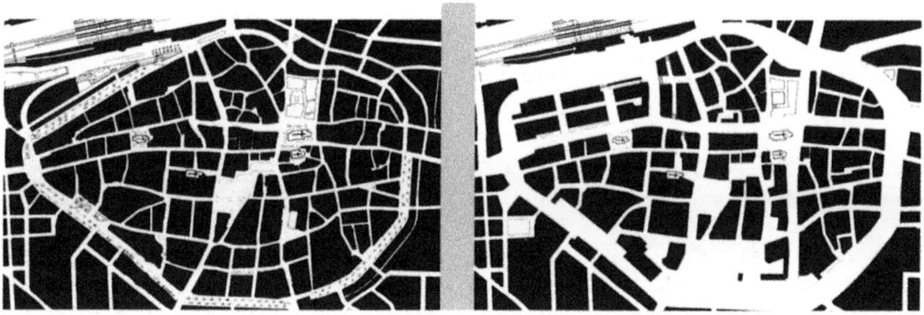

The planning policy of Dortmund took less account of historic structure. The city's design focused on wide, straight avenues. Only the churches kept their place in the new city, linking it to its mediaeval history. Such far-reaching interventions in the city's layout had been seen in neither the Wilhelmine era nor in 1920. The top priority of a car-friendly city initiated a massive reformatting of the urban soundscape.[29]

29 Heine-Hippler, "Will Schwaz", 62.

Figure 7: New perspectives for cars; only the churches kept their place in the new concept, photo: Gerhard Fleitmann

The massive devastation and the influx of displaced people, refugees and expellees looking for jobs in the Ruhr presented the local authorities with immense problems of urban development. Five years after the war, 500,000 people were living in Dortmund again.

Figure 8: New sounds through new building materials in the new consumer city centre, photo: Gerhard Fleitmann

In 1953, the city had as many inhabitants as ever, but the pre-war housing stock had not yet been reached.[30] There was an urgent need for housing. Large-scale terraced housing estates were built around the industrial spaces, the model espoused being that of the "gegliederte und aufgelockerte Stadt" ("dispersed and structured city").[31] There arose a new soundscape, which is charmingly documented in the 1952 educational film *The Slagheap Playground.* At the end of

30 Heine-Hippler and Trocka-Hülsken, "Wohnungsbau".

31 See illustrations in Framke, *Das neue Dortmund*, 90–94; see also Fürst, Himmelbach and Potz, *Leitbilder*, 42–45.

the film, the young protagonist, a football player, moves to a new suburban settlement.[32]

Figure 9: Dortmund-Scharnhorst large-scale housing estate, following the German urban planning model of the "gegliederte und aufgelockerte Stadt" ("dispersed and structured city"), photo: Walter Moog, DOGEWO

In 1945, the time of the air raid sirens was not yet over. The British authorities used them to announce a dusk-till-dawn curfew. In May 1945, it lasted from 9 p.m. to 5 a.m. The sirens sounded for one minute, fifteen minutes before and at the beginning of the curfews.[33] The Federal Republic of Germany's civil defence continued to make use of it to prepare the population for a nuclear first strike. Thus it became also one of the acoustic signatures of the Cold War.

The Dortmund war chronicle of 1945 noted the following about the British information policy: "Every Thursday at 1 p.m., official messages were spread through speakers on Hansa Square, in Hörde and in Aplerbeck".[34] For 25 May 1945, the chronicle writes about the resumption of some train connections. From

32 Westfälisches Landesmedienzentrum, *Der Platz an der Halde. Ein Film von Frank Leberecht und Herbert Fischer* 1952/53, 35 mm, 52 Min.

33 See document in *Heimat Dortmund*, 1 (2015), 69.

34 Ibid., 68.

15 June, the tram re-linked the Hörde suburb to the little town of Schwerte on the River Ruhr.[35] The British authorities announced the opening of cinemas for Monday, 30 July. One hour before the curfew started, the cinemas had to close, so that the audience could find their way home.[36]

Figure 10: Announcement: "The cinemas will open again. Blockbusters first", Ruhr Zeitung, *27 July 1945, Dortmund municipal archives*

Ruhr Zeitung

HERAUSGEBER: DIE BRITISCHE BESATZUNGSBEHÖRDE

Ausgangsbeschränkung

29. Juli — 11. August

von 21.45 bis 05.15

Ab 1. August werden die Zeiten des Ausgehverbots in allen Städten und Dörfern, die über Fliegersirenen verfügen, täglich durch die Sirene angekündigt.

15 Minuten vor Beginn der Ausgehverbotszeit wird Fliegeralarm gegeben.

Bei Beginn der Ausgehverbotszeit wird das Signal „Gefahr vorüber" gegeben.

Nr. 12 | Freitag, den 27. Juli 1945 | Preis 20 Pf

Montag eröffnen die Kinos

Erst kommen Spielfilme zur Aufführung

In der nördlichen Rheinprovinz und in Westfalen ist die Eröffnung von 100 Lichtspielhäusern für die nächsten beiden Wochen vorgesehen. 120 Kinos in der nördlichen Rheinprovinz und etwa die gleiche Anzahl in Westfalen beginnen bereits am Montag, dem 30. Juli, mit ihren Vorstellungen.

In der ersten Zeit wird in jedem Kino nur ein Film gezeigt werden, und zwar ein Spielfilm. Die Militärregierung wird örtlich die Anzahl und Zeiten der Vorstellungen bestimmen. Ueberall werden die Lichtspielhäuser eine Stunde vor Eintritt des Ausgangsverbots geschlossen, so daß das Publikum genügend Zeit findet, nach Hause zu gehen.

Britische und amerikanische später

Zunächst werden nur deutsche Filme vorgeführt werden, es ist aber zu erwarten, daß binnen kurzem britische und amerikanische Filme beschafft werden können.

Die Verzögerung der Wiedereröffnung der Lichtspielhäuser, die seit langem vom Publikum erwartet wurde, ist darauf zurückzuführen, daß die Wiederinstandsetzungsarbeiten in einem Gebiet, das so [illegible] Kriegsschäden davongetragen hat wie das Rheinland und Westfalen, eine besonders schwierige Aufgabe darstellten.

In September 1945, there came a sound event of outstanding importance, marking the start of the industrial reconstruction: "On 26 September 1945, at Hörder Hüttenverein [steel plant], the first tapping of pig iron since the Second World War", notes the annalist Luise von Winterfeld.[37] On 11 November 1945, the military government decreed a two-minute silence to commemorate the fallen soldiers of the two world wars. Von Winterfeld laconically notes: "two minutes of silence, halted traffic, etc. at 12 p.m. In Dortmund, you didn't notice anything at all".[38]

35 Ibid., 68.
36 *Ruhr Zeitung*, 27 July 1945; see document in *Heimat Dortmund*, 1 (2015), 69
37 *Heimat Dortmund*, 1 (2015), 70.
38 Ibid., 72.

The world is waiting for the sunrise

While most inhabitants of the Ruhr were proudly listening to the sound of the reconstruction, dangerous new sounds came from pubs, music halls, restaurants and cinemas. During the post-war period, up to the Währungsreform (currency reform), the restaurant "Zur Postkutsche" in the Aplerbeck district became an extraordinary music venue. The Marktplatz Aplerbeck tram stop had more passengers than ever before. On Saturdays and Sundays, the trams offered extra tours to Aplerbeck. They transported amusement seekers, who loved to dance with jazz bands like Walter Hanke, Karl Scheel, Charly Little or Joe Wick.[39]

Figure 11: Hall of the "Postkutsche" in Dortmund-Aplerbeck, photo: private collection

These bands played the hottest jazz ever. In the audience, as well as on stage, there were young guys who had just been exposed to jazz during the Nazi era. Many of them had their first contact with authentic jazz in the Nazi military. Pit Buschmann, a talented jazz guitarist, heard his first authentic jazz record at the Flak in Dortmund-Wellinghofen. For him, jazz symbolised a new era. This was the music of the liberators, a music signifying humanity, tolerance, freedom and democracy. For most Germans, however, it was the music of the occupiers: "American" mass culture. For German bourgeois critics, jazz – as music as well

39 Schmidt, Müller and Ortmann, *Jazz in Dortmund*, 54

as a lifestyle – signalled the collapse of the established order, the mixing of races and the dissolution of gender roles in dance and entertainment.

Dortmund musicians who started their career as jazz musicians at that time remembered tunes of key significance to them. "Bonny" Schüten (today nobody knows his real name![40]), founder and banjo player of the Dortmund band Dark Town Stompers, listened to his first jazz tune on Radio Munich, broadcast by the American military government. It was "The World is Waiting for the Sunrise" by Benny Goodman. Other musicians cited Goodman's "Moonglow" as their crucial experience: "This music is my music, this music I want to play!"[41]

Few artists could have articulated the dawn of the new age so poignantly as Benny Goodman. Again and again, the Nazi propaganda had shown his hands on the clarinet flaps for its racist defamation, titled "Verbrecherhände" ("hands of a criminal").[42] "Glen" Buschmann – later director of the Dortmund municipal music school – got a job in Joe Wick's band, one nucleus of the developing big band scene in West Germany. Thanks to the financial support of the British Army Welfare Service, Joe Wick's orchestra became the leading West German big band. Jazz lovers praised it as the best band the German jazz scene had to offer at that time. As Joachim Ernst Behrend, known as the "German Jazz Pope", emphasised: "Hearing the first Brunswick records of Joe Wick, I was sure this had to be an American band, with black musicians. It was inconceivable that in 1947 a German band could play such authentic jazz music".[43]

40 Young jazz musicians chose an American first name like "Glen", "Pit" or "Bonny" for their new identity.

41 Schmidt, Müller and Ortmann, *Jazz in Dortmund*, 57.

42 See headline and article in *Illustrierter Beobachter, Folge*, 1944/26.

43 Schmidt, Müller and Ortmann, *Jazz in Dortmund*, 59.

Figure 12: The Joe Wick Orchestra, June 1948; photo: Archiv Horst Lange

Yet most contemporaries listened to something different. The narratives of concert reviews illustrate how strongly the racist ideology of the middle-class national conservatives and the Nazis still held people in thrall: "Joe Wick is for drugs", the *Westdeutsche Allgemeine Zeitung* headlined after a concert in Dortmund.[44] And the *Westfälisches Tageblatt* wrote: "Players and audiences are mesmerised by syncopation! The bassist hits his bass to the point of ecstasy. The trumpet player blows a double high C with rolling eyes. The massed audience beats the time. Are they betraying Bach?"[45]

Interestingly, in contemporary concert reviews of the post-war era, the trumpeter is always playing a "C" with "rolling eyes", a racist image of the childlike, primitive "Nigger". You can find this image in films and pictures, even repeated by the trumpeter Louis Armstrong himself. It refers to the Blackface Minstrelsies, a form of entertainment in the US that was developed in the early nineteenth century. The show lampooned black people as dim-witted, happy-go-lucky children. Associated with ecstatic bodies, percussion instruments and a high tempo, this narrative evokes the primitive energy of jazz music. It highlights the perils of the "abendländische Kultur", the corroding of white maleness by the "cultural

44 Ibid., 59.
45 Ibid., 60.

feminism" of American mass culture – a favourite bourgeois German discourse disdaining mass culture since the Wilhelmine era.

Indeed, in the main, newspapers did not inform people about the music of the Joe Wick Orchestra, but pointed out fears about national identity within the context of German capitulation and American consumer culture. This thesis is confirmed by concert reviews in newspapers from southern Germany, which sound almost identical to western ones from the Ruhr. Joe Wick, born in 1916 in the Rhineland as Josef Wick, clearly participated in this discussion with his music performances. It can be read between the lines of a concert review published in the *Westfälische Rundschau*, a major newspaper in the Ruhr at that time. Joe Wick's band had played in the newly reopened "Assauer Lichtspiele" in the north of Dortmund. The newspaper wrote: "A little later, during the 'Chattanooga Choo-Choo', the seventeen musicians of the rhythm section, one after the other, packed their instruments and left the stage. In front of the stage, Joe shouldered his violin. He couldn't stop the band dissonantly playing 'Old Comrades' as jazz, while absurdly performing a kind of parade".[46]

The military march "Old Comrades" [Alte Kameraden] represented (and still represents!) German militarism at its best: its commando structures, marching order, goose steps and strong white German maleness. Today, this sophisticated review reflects a critical anti-military music performance and an explicit political message, staged by the Joe Wick Orchestra: should Germany at last be on the path to coolness, casualness and a democratic attitude – after all the devastating militarism, nationalism, masculinism and racism?

How did the post-war period sound to contemporaries? Every night, people woke up screaming, wartime experiences still hammering in their heads. Wolfgang Borchert, writer of famous short stories like "The Man Outside", implies: "For us, our sleep is filled with battle. Our night is filled with battle noise in its dream-death […]. Who will write new laws of harmony for us? We no longer need well-tempered pianos. We ourselves are too dissonant. […] We no longer need any still-lives. Our life is loud".[47]

---

46 *Westdeutsche Allgemeine Zeitung*, 23 October 1948, *Westdeutsches Tageblatt*, 16/23 October 1948; quoted in Schmidt, Müller and Ortmann, *Jazz in Dortmund*, 60.

47 "Wer schreibt für uns eine neue Harmonielehre? Wir brauchen keine wohltemperierten Klaviere mehr. Wir selbst sind zuviel Dissonanz. […] Wir brauchen keine Stillleben mehr. Unser Leben ist laut". Borchert, "Das ist unser Manifest", 113.

Conclusion

People learned to gain their bearings in space and time thanks to battle sounds, signals and sirens. But in 1945, the time of the sirens was not yet over. The Federal Republic of Germany continued using air raid sirens. For many, nights and dreams were still filled with the noises of war. So they became also acoustic signatures of the Cold War. Epistemologically, the opposition between "roaring war" and "silent peace" does not work. That fixed polarisation does not apply to reconstructing the urban soundscapes of "1945", with the public and private dimensions, inner and outer sounds. We need simultaneous thinking about the non-simultaneous, about overlapping soundscapes. We need case studies of examples, situations, details and fragments. We need case studies that are aware of anthropological and technical, as well as spatial, references.

Bibliography

Abeck, Susanne, and Schmidt, Uta C. (eds), *Stulle mit Margarine und Zucker. Heimat Ruhrgebiet* (Essen: Klartext, 2015).

Blank, Ralf, "Kriegsende in Dortmund – die zerstörte Stadt", *Heimat Dortmund*, 1 (2015), 3–12.

– "Kriegsendphase und 'Heimatfront' in Westfalen", *Westfälische Forschungen*, 5 (2005), 361–421.

Böhme, Gernot, *Anmutungen. Über das Ästhetische* (Ostfildern: Ed Tertium 1998).

– *Atmosphäre. Essays zur neuen Ästhetik* (Berlin: Suhrkamp 2013).

Borchert, Wolfgang, "Das ist unser Manifest", in *Draußen vor der Tür* (Reinbek bei Hamburg: Rowohlt 1973).

Bull Michael, and Back, Les (eds), *The Auditory Culture Reader* (Oxford: Berg, 2003).

Chartier, Roger, "Kulturgeschichte zwischen Repräsentationen und Praktiken. Einleitung", in *Die unvollendete Vergangenheit. Geschichte und die Macht der Weltauslegung* (Berlin: Wagenbach 1989), 7–20.

– "Die unvollendete Vergangenheit. Beziehungen zwischen Philosophie und Geschichte", in *Die unvollendete Vergangenheit. Geschichte und die Macht der Weltauslegung* (Berlin: Wagenbach 1989), 21–36, first published in *Philosophie et histoire*: actes du séminaire tenu à Paris, au Centre Georges-Pompidou, les 23, 24 et 25 avril 1986, organisé par l'Espace de séminaire du Centre Georges-Pompidou et l'Espace international de recherche philosophique et interdisciplinaire (Paris: Centre Georges-Pompidou, 1987), 115–135.

Corbin, Alain, *Les cloches de la terre: paysage sonore et culture sensible dans les campagnes au XIXe siècle* (Paris: Michel, 1994).

Feiereisen, Florence, and Hill, Alexandra Merly (eds), *Germany in the Loud Twentieth Century* (New York, NY: Oxford University Press, 2012).

Fürst, Franz, Himmelbach, Ursus, and Potz, Petra, *Leitbilder der räumlichen Stadtentwicklung im 20. Jahrhundert – Wege zur Nachhaltigkeit?*, Berichte aus dem Institut für Raumplanung, 41 (Dortmund: Institut für Raumplanung, 1999).

Geertz, Cliffort, "Thick description: toward an interpretive theory of culture", in *The Interpretation of Cultures: Selected Essays* (New York: Basic Books, 2000), 3–30.

Grontzki, Nina, Niewerth, Gerd, Norbisrath, Gudrun, and Potthoff, Rolf (eds), *Feuersturm und Hungerwinter. Zeitzeugen erinnern sich an Krieg und Wiederaufbau* (Essen: Klartext 2007).

Günther, Dorothea, http://www.dhm.de/lemo/zeitzeugen/dorothea-günther-der-krieg-193940, accessed 29 March 2017.

Heine-Hippler, Bettina, "Will Schwarz und die Dortmunder Gespräche", in Gisela Framke (ed.), *Das neue Dortmund. Planen, Bauen, Wohnen in den fünfziger Jahren* (Dortmund: Kettler, 2002), 57–67.

–, and Trocka-Hülsken, Ingrid, "Wohnungsbau in der Nachkriegszeit", in Gisela Framke (ed.), *Das neue Dortmund. Planen, Bauen, Wohnen in den fünfziger Jahren* (Dortmund: Kettler, 2002), 80–94.

Krause, Joseph P., http://ww2today.com/6-november-1944-black-monday-for-gelsenkirchen-as-raf-bombers-return, accessed 29 March 2015.

Leberecht, Frank, and Fischer, Herbert (directors), *Der Platz an der Halde* (film), 1952/53, Westfälisches Landesmedienzentrum, 35 mm, 52 Min.

Missfelder, Jan-Friedrich, "Period Ear. Perspektiven einer Klanggeschichte", *Geschichte und Gesellschaft*, 38 (2012), 21–47.

Moeller, Robert G., "Remembering the war in a nation of victims. West German pasts in the 1950s", in Hanna Schissler (ed.), *The Miracle Years. A Cultural History of West Germany 1949–1968* (Princeton, NJ: Princeton University Press, 2001), 83–109.

Morat, Daniel, "Sound Studies – Sound Histories. Zur Frage nach dem Klang in der Geschichtswissenschaft und der Geschichte in der Klangwissenschaft", *kunsttexte.de/Auditive Perspektiven*, 4 (2010), 1–7, http://www.kunsttexte.de/index.php?id=711&idartikel=37458&ausgabe=37455&zu=907&L=0, accessed 29 March 2017.

Schafer, Murray R., *The Tuning of the World* (New York: Alfred A. Knopf, 1977).

Schmidt, Uta C., "Radioaneignung", in Inge Marßolek and Adelheid von Saldern (eds), *Zuhören und Gehörtwerden I: Radio im Nationalsozialismus. Zwischen Lenkung und Ablenkung* (Tübingen: Discord, 1998), 243–355.

–, Müller, Andreas, and Ortmann, Richard, *Jazz in Dortmund. Hot-Modern-Free-New* (Essen: Klartext 2004).

Schneider, Angela, and Kiefer, Anselm, *Anselm Kiefer* (Berlin: Nicolaische Verlagsbuchhandlung Beuermann, 1991).

Urmersbach, Karlheinz, http://www.gelsenzentrum.de/bombenangriff_1944.htm, accessed 29 March 2017.

Weinauge, Willy (ed.), *Von der toten zur lebendigen Stadt. Fünf Jahre Wiederaufbau in Dortmund* (Dortmund: Dortmunder Zeitschrift, Schinker, 1951).

Wieden, Lotta, "Maikäfer, flieg! Die meisten von uns kennen es noch: das Maikäfer-Lied. Aber warum eigentlich? Warum ist diese alte Kinderweise noch nicht in Vergessenheit geraten? Unsere Autorin hat sich auf eine Spurensuche quer durch Deutschland gemacht", *Frankfurter Allgemeine Sonntagszeitung*, 5 April 2015, 43–44, http://www.faz.net/aktuell/gesellschaft/altes-kinderlied-maikaefer-flieg-13522509.html?printPagedArticle=true, accessed 29 March 2017.

Katarzyna Naliwajek-Mazurek

Institute of Musicology, University of Warsaw

The sounds of Warsaw in 1945: Witness accounts

It may well be only through testimony that we can really write the history of such a cataclysm.
Dori Laub[1]

Abstract: Testimonies of Warsaw at the turn of 1945 include auditory memories related to life in the occupied city during and after the Warsaw Uprising. Traumatic experiences are intertwined with sonic experiences, such as the sounds of warfare. The accounts also contain trauma-related musical memories. Interviews with witnesses gathered by the author and from other sources are considered from the point of view of traumatic memory to see what types of sounds were remembered, how they were recalled and whether some types of auditory experience could play a special role in posttraumatic responses, such as the creative reintegration of the self.

Keywords: traumatic memory, warfare sounds, trauma-related sounds

In the narratives of witnesses who lived in and around Warsaw at the turn of 1945, auditory experiences and representations appear as part of a complex network of memory threads. The testimonies are narrated as a more or less continuous series of memories, linking 17 January 1945 (the day when the city was liberated from the Nazis) with reminiscences of the previous six months, particularly the intense and traumatic experiences from the time of the Warsaw Uprising of August and September 1944. Given the short period, these events seem naturally connected in time, yet their continuity was actually punctuated by several disrupting moments that drastically altered the witnesses' living conditions. Already existing in perilous situations, they went through a series of extremely traumatic events, such as the death of family members, the loss of their home, bombardments and other German military and genocidal actions against civilians, forced expulsion from the city and then a return to a Warsaw of ruins and graves. Those experiences make them representatives of a "traumatized population", defined as "those who have been exposed to severe life threat, either to the self or to significant others; those who have been

1 Laub, "From speechlessness to narrative", 264.

exposed to violence or the gruesome and mutilating deaths of others; or those who have experienced a 'death encounter'".[2] The testimonies analysed below, gathered from survivors interviewed mainly by the author of this chapter, involve sound memories within the context of such traumatic experiences.

Although the research project within which these interviews were gathered was historically rather than psychologically orientated, it seems relevant to consider these trauma-related sound memories from the perspective of traumatic memory studies. The introduction of these psychological theories makes it possible to explain some aspects of the mechanisms involved in remembering sounds in traumatic contexts.

Traumatic memory and sounds

According to *The Oxford Handbook of Memory*, the three most often mentioned mechanisms traditionally linked with the notion of traumatic memory are repression, dissociative tendencies and the supposedly purely sensory character of such memories, yet the authors of the *Handbook* note that the crucial question as to whether traumatic memory involves any special mechanisms has been "a topic of marked controversy".[3] They quote a study comparing traumatic and nontraumatic memories which demonstrated that trauma victims "reported that they initially remembered the traumatic event in the form of somatosensory or emotional flashback experiences, and a narrative memory began to emerge only later. In contrast, nontraumatic events were recalled as narratives without sensory components".[4] However, as other researchers have argued that no such unique mechanisms exist, the authors suggest that trauma-related processes might be "apt to be associated with other types of memories",[5] hypothesising that "traumatic memories can be remembered with excessive vividness or not recalled at all, depending (perhaps) on both the pattern of cognitive/neural activity of encoding and the nature of the environmental and experiential cues that are encountered later".[6] Notwithstanding all the controversies, further research into traumatic memory and emotions in survivors' testimonies has brought explanations that are useful for the study of sound memory and its relevance in the context of war trauma. Robert Kraft,

2 Raphael and Dobson, "Acute posttraumatic interventions", 141.
3 Tulving and Craik, *The Oxford Handbook of Memory*, 386.
4 Van der Kolk and Fisler. *Dissociation*, 505–525. Quoted in Tulving and Craik, *The Oxford Handbook of Memory*, 387.
5 Tulving and Craik, *The Oxford Handbook of Memory*, 388.
6 Ibid.

in his study on emotional memory in Holocaust survivors, stated that traumatic memory as analysed in oral testimony is characterised first of all by its episodic structure. The episode, a small story, is an encapsulated "primary unit of remembering", usually in the form of "a narrative, although it can be impressionistic, and the episode may be placed within a precisely described spatial layout. The basic elements of the episode are vivid perceptual images, deeply felt emotions, and physiological experiences". They are "connected to one another by thin strands that are either chronological and geographic or thematic or rhetorical".

The second crucial feature of traumatic memory is the duality of memorial representation: "memory represents the original phenomenal experience in the form of images, emotions, and bodily sensations"; "original memories are then integrated into a narrative, episodic construction of events".[7]

These two modes of remembering are described in other terms as "deep memory" versus "external" or "intellectual" memory,[8] "deep memory" versus "common memory", or "situationally accessible memory" versus "verbally accessible memory".[9] Kraft defined this duality as core memory and narrative memory, explaining it as follows:

> Core memory is the representation of the original phenomenal experience in the form of perceptual, emotional, and physiological experience: visual images, sounds, smells, tastes, emotions, and bodily sensations. [...] Narrative memory is constructed from the images in core memory, shaped in accordance with narrative conventions and conveyed primarily in language. The word "narrative" characterizes how people structure episodes in personal memory for the purpose of thinking about oneself in silent remembering and communicating with others to tell the events of the past [...].[10]

In light of Kraft's theory, the remembrance of sounds as elements of phenomenal experience pertaining to the core memory is then retold by means of narrative memory, which is constructed, for example, according to the chronological order of events. This mechanism explains to a certain extent how the effort of narrative memory helps an individual to regain rationality through the reconstruction of continuity. It is also a more or less conscious method employed to counteract posttraumatic amnesia. Remembered sounds thus constitute elements

7 Kraft, "Emotional memory", 352.

8 See Anne Whitehead's discussion of Charlotte Delbo's "influential description of traumatic memory" in her *Days and Memory*. Whitehead, *Memory*, 117 ff.

9 Similar discernment is present in "fuzzy-trace theory": "verbatim memory is the original representation of the events; gist memory structures and summarizes verbatim memory and is in linguistic form". Kraft, "Emotional memory", 354.

10 Ibid., 353.

of multi-layered landscapes of the survivors' memory. In some cases, the witnesses interviewed by the author consider sounds as crucial components of a narrative (see the testimony of Halina Paszkowska-Turska); in other instances, they see them as secondary or as an acoustic background. In the case of most traumatic recollections, sound may be not mentioned at all or become a decisive element in the experience of moments permeated by the trauma of imminent death.

Although the hypothesis that nontraumatic events are "recalled as narratives without sensory components"[11] whilst traumatic memory contains such sensory material cannot be either sufficiently sustained or rejected through the analysis of the testimonies presented below, as they were not part of a systematic psychological study, post-war recollections of 1945 Warsaw seem to belong to a different type of internal narrative than war-time memories. When asked about specific sound memories, witnesses mention, for example, the sounds of trams when they reappeared in the right-bank Praga district of the city,[12] the return of saw players to the ruined streets or the sounds of prostitutes hailing clients,[13] but they tend to concentrate on musical rather than sound events. For this special group of witnesses, who were interviewed by the author of this chapter for the purpose of reconstructing the history of musical life, and also for those asked specifically about sounds in Warsaw around 1945, music is one of most important elements in their lives. Thus concerts and the gradually reconstructed cultural life became crucial components (more meaningful than the reappearance of the sound of trams) of the gradual return to "normal" life, free of the previously omnipresent trauma-inducing sounds of warfare.[14]

Posttraumatic responses and sound memory

The study of survivors' narratives from the point of view of posttraumatic stress disorder (PTSD) as analysed by theories informed by experimental psychology and information processing theory could further explain the process of memory

11 Van der Kolk and Fisler, *Dissociation*.

12 The testimony of musicologist Andrzej Spóz, 2016.

13 The testimony of art historian Andrzej Olszewski, 2017.

14 Carolyn Birdsall mentions the distressing influence of a warfare soundscape: "The overpowering sounds of modern warfare – with shells, guns and artillery – were produced at an intensity unlike anything experienced or imagined before. This soundscape, dominated by the 'technologized sonority' of warfare, was the basis for distress, and often lasting trauma, for those subjected to its relentless noise and overwhelming of the self". Birdsall, *Nazi Soundscapes*, 15.

structuring, which seems revelatory also in relation to sound memory. Here is the definition of trauma given by the American Psychiatric Association:

> direct personal experience of an event that involves actual or threatened death or serious injury, or other threat to one's physical integrity; or witnessing an event that involves death, injury, or a threat to the physical integrity of another person; or learning about unexpected or violent death, serious harm, or threat of death or injury experienced by a family member or other close associate [...].[15]

It should be stressed, however, that as early as 1997 the validity of the PTSD concept was questioned and began to be viewed as a cultural product from anthropological stances,[16] and many researchers, "including anthropologists and cross-cultural psychiatrists, such as Arthur Kleinman, have strongly criticized the labelling of acute responses to traumatic violence as a mental disorder, arguing that this represents a pathologization and professionalization of normal forms of human suffering, a form of globalized medicalization of human responses to disaster and violence".[17] PTSD is being replaced with more nuanced concepts, such as "complex PTSD", which "may result from 'a range of adult experiences, including war, civil conflict, torture, and other experiences involving pervasive loss of control over aversive consequences'".[18] In social contexts, terms such as "collective or mass trauma", "community disaster" and "cultural trauma" are useful.[19] Posttraumatic responses might involve processes that can operate positively to counteract the debilitating effects of trauma. Such mechanisms as "resilience", described by Roni Berger as "the ability to withstand and respond to a stressor event, traumatic exposure, or crisis in a constructive way and bounce back from adverse experiences and circumstances quickly and effectively" and "posttraumatic growth" ("benefits (growth) or positive changes that can be gained from struggling with a highly stressful event include strengthening of individuals, families and communities, discovering abilities and talents not previously recognized") seem particularly helpful.[20] On the intrapersonal level, structuring mechanisms that occur in the processes of traumatic memory recovery may play such a positive role. This hypothesis was proved by a psychological test conducted by Dori Laub. His study brought evidence of the profound healing power of the process

15 *Diagnostic and Statistical Manual*, 274. Quoted in Berger, *Stress*, 8.

16 Cf. Allan Young's ground-breaking book *The Harmony of Illusions*.

17 Good and Hinton (eds.), *Culture and PTSD*, 8.

18 Quoted after Good and Hinton (eds.), *Culture and PTSD*, 6.

19 Berger, *Stress*, 9–12.

20 Berger, *Stress*, 13.

of verbalising testimonies: "the scores of trauma-induced symptoms (in particular the withdrawal cluster of PTSD) had been reduced by close to 50 percent in comparison with a control group".[21] The process of healing traumatic memory symptoms related to PTSD consists partly in "the accurate recall of traumatic stressors and experiences; this includes learning the relevance and significance of amnesia, or gaps in memory and the chronology of experience, in order to reconstruct a relatively complete memory for the traumatic event".[22] Thus the chronological structuring of memory gains an important therapeutic dimension. This is further validated by another important goal of treatment, which is to "restore integration of the self and facilitate the transformation of cognitive schema from victim identity to survivor identity with a sense of coherence and continuity in time, space, and culture".[23] On the personal level of the survivors, the scale of reactions accompanying the reformulation of trauma range from self-destructive to creative tendencies. The restoration of the self may or may not occur during this "protean task", and the "incongruities within the self may motivate creativity, drama, or self-destruction".[24] The reasons for these differently evolving processes are explained thus:

> In healing, some survivors experience a restoration of self-sameness and continuity. For other survivors, a void exists between two psychological realities created by the advent of trauma. This discontinuity is typically experienced as the "old self" and its replacement after trauma by a "different self".[25]

Sound memories may become the "fuel" of such positive processes, which can be observed in the biographies of interviewed witnesses who lived through such challenges resulting in their experience of cataclysm.

It might be argued that this hypothesis is sustained by a qualitative psychological study that examined the roles of music for Holocaust survivors who were musicians.[26] According to the authors' description, six out of seven participants were amateur musicians, with just one professional. The findings showed various roles played by music in the participants' survival of the Holocaust, such as granting identity or coping strategies (distraction, venting), helping others, and finally physical survival. Music was also crucial at the preliminary stage

---

21 Laub, "From speechlessness to narrative", 259.

22 Wilson, *An Overview*, 78–9.

23 Ibid.

24 Ibid.

25 Ibid.

26 Fisher and Gilboa, "The roles of music", 1221–1239.

when the Holocaust began, at the time of instability and chaos, when the Jews were denied their civil rights and music played a crucial role in the witnesses' attempts to maintain their routines. It also helped them rehabilitate and return to normal life after the Holocaust ended. It constituted a major factor for their identity formation before the Holocaust and in identity re-formation after the Holocaust.[27]

Chronologies of memory

Most of the witnesses to the war-time and immediate post-war realities in Warsaw whose testimonies are quoted below were interviewed by the author between 2014 and 2017. The video and audio recordings took place at their homes and were accompanied in some cases by further meetings and conversations. A few of them were made specifically for the sake of this study, when questions about sound memories were asked. In all cases, however, these outstanding individuals were chosen because of their particular musical and auditory memories. All of them are linked with music in various ways; if not always professional musicians, they are music lovers, particularly sensitive to auditory elements.

The fact that the interviewees have lived longer than the majority of their peers seems particularly meaningful in the light of the above-quoted theories, especially when we consider that all of them are highly creative individuals. Their professional fields include sound engineering (Halina Paszkowska-Turska), theatre (Witold Sadowy), medicine (Halina Szpilman), art history (Andrzej Olszewski), music composition (Włodzimierz Kotoński), instrument playing (Jadwiga Kotońska), musicology (Władysław Malinowski, Andrzej Spóz) and journalism (Wacław Gluth-Nowowiejski). This sole fact may initially sustain the hypothesis that traumatic memory (and trauma-related sound memories) can be reworked as part of a redefined creative self. The analysis of the testimonies investigates the character and function of sound memories in these narratives and the possible role could they play in posttraumatic reformulations.

In the testimonies not collected by the author, musical or auditory memories were incidental to the recollections. These sources are twofold: transcripts of interviews made for the Museum of the Warsaw Uprising and memoirs written by witnesses themselves. The former reveal which sounds are most often mentioned by survivors of the Warsaw Uprising and which are described as the most traumatic, allowing us to advance a tentative typology of trauma-related sounds. The latter

27 Ibid., 1227 ff.

provide information on the role of music in the situation of war-time trauma. The interspersed source material from interviews conducted by the author brings more detailed, individual perspectives. Thus these heterogeneous sources are analysed from the point of view of auditory (and musical) memories cited by the witnesses.

These memories cover such markedly different periods in the history of Warsaw as the occupation, up to 1 August 1944, when the city still functioned in many respects (e.g. public transport), the destruction of buildings and the mass murder of inhabitants during and after the Warsaw Uprising, the transitional time of expulsion and displacement and the return to the ruined city after 17 January 1945, when Warsaw was liberated from the Wehrmacht and taken under Soviet control.

The ordeal of Varsovians transported to camps was often prolonged until May 1945, whereas those who managed to escape the transports or fled the city before or after the Warsaw Uprising were able to return to the city already in January 1945. Then there were those rare cases of residents who never left the city, hiding in houses and among the ruins. Some of them survived, like the pianist Władysław Szpilman. Others were murdered by German soldiers leaving the city, such as another outstanding pianist of the inter-war era, Róża Etkin-Moszkowska, also of Jewish origins.[28] These chronologies become more complicated when we consider that while the left bank of the city was subject to the atrocities and cruelty of German and ROA (Russian Liberation Army) soldiers, the right bank (Praga) was captured by the Red Army and the Polish Army under the command of generals Zygmunt Berling and Wojciech Bewziuk between 10 and 15 September 1944. Already on 15 September 1944, Colonel Marian Spychalski, newly appointed mayor of right-bank Warsaw by the Polish Committee of National Liberation (Polski Komitet Wyzwolenia Narodowego, PKWN),[29] issued a call to "people of culture" to begin work. The next day, the conductor Jerzy Wasiak, who had earlier worked in German-approved theatres and conducted the popular Radiana Choir, showed up at the Culture and Art Bureau set up in Praga; he would soon organise the City of Warsaw Symphony Orchestra. So cultural life, albeit very limited, was being organised on the right bank of the Vistula, whereas the left bank was being mercilessly destroyed by German forces, with civilians falling victim to executions

28 This colleague of Szpilman was killed (according to Szpilman's testimony) on 16 January 1945 with her husband, the eminent architect and sculptor Ryszard Moszkowski, in the Żoliborz district, where she had been hidden by friends during the whole occupation. See Dybowski, *Słownik*, 161.

29 This was a puppet provisional government of Poland, controlled by the Soviet Union, proclaimed on 22 July 1944, in opposition to the Polish government in exile. See Snyder *Bloodlands*, 96.

and murder.[30] Taking all these factors into consideration, it seems ineffective to analyse testimonies beginning strictly from 1 January 1945; such an approach would be artificial and prevent us from grasping the unfolding drama of events described by the witnesses and the transformations of the soundscape in Warsaw around the turn of 1945.

Trauma-related sounds

Different trauma-related sounds as remembered by witnesses are described below, ordered according to types that reflect their character and their psychological effect on the witnesses. The ordering is chronological wherever possible and partly reflects the order in witnesses' accounts.

Sounds instilling anxiety, fear and panic linked with life-threatening conditions were of a different character during the three main war-time periods mentioned above.

In testimonies from the time before the Uprising, several witnesses mention highly traumatic sounds related to arrests and killings of Warsaw inhabitants. The "soundscape itineraries" of Gestapo arrest scenes are omnipresent in survivors' accounts. They usually contain similar auditory elements and chronologies: a car stopping in front of the house in the night, military boots heard on the stairs, banging at the door, shouting in German. These actually heard sounds are combined in witnesses' memories with reminiscence of intensely imagined sounds that accompanied apprehension about arrest, which in some cases was not directly verbalised, but expressed through a narration of subsequent sounds. In cases when the Gestapo entered witnesses' homes, the memory preserves also the actually experienced atrocities committed.

Ludwik Erhardt (b. 1934), who after the war became a musicologist and music critic,[31] remembered in his written memories a sound that signalled a dramatic moment in his family's life during the occupation:

> Another scene, present in my memory to this day: a violent ring at the door, uniforms, my mother white as a sheet, a search. I understood the meaning of this scene much later. On Okólnik [Street], in the small room, "Mr Zygmunt" lived with us, a close friend from pre-war times, a senior officer from Vilnius region, who relinquished his uniform and changed his name. Someone gave him up and he was arrested on the street. If the search

30 These facts are quoted in Mościcki, *Teatry*, 426–431.

31 He studied musicology at Warsaw University (1952–1956), was editor (from 1957) then editor-in-chief (1971–2008) of the journal *Ruch Muzyczny*, and wrote books on Schumann, Brahms, Stravinsky and Penderecki, as well as several others.

> of his room had been more meticulous, we would have been lost. There were weapons and important documents under the floor. Mum knew that. Major Zygmunt Bohdanowicz was a member of the AK [Home Army] command. He died in Auschwitz.[32]

The hidden violence of this scene (he was then a child unaware of its meaning) is imprinted in Erhardt's memory and expressed by the violence of the sound of the doorbell. Like several other survivors, often children at the time, he also remembered the sights and sounds that accompanied a public execution:

> One's memory preserves different images from childhood, like photographs on a computer. With some of them, I would like to press "delete". I usually walked between Królewska and Okólnik [streets], so along Krakowskie Przedmieście, Nowy Świat and Ordynacka. Sometimes it was a difficult route. Once at the corner of Nowy Świat and Świętokrzyska, I found myself in a crowd of people standing in silence. I was only nine, but I did not have to ask anybody for an explanation anymore. There were captives being shot in front of halted passers-by. I saw it. There is a plaque to this day.[33]

The strongest recollection in this case was the sound of silence after the atrocity of the victims' execution – silence that could express such different emotions as horror and fear, supressed anger and helplessness, but also the intention to commemorate and mourn the victims.

The traumatic sonority of the German language is mentioned in several testimonies, as it is directly associated with atrocities committed by the Nazis. In the testimony of Barbara Hanna Sielicka (1929–2008, pseudonym "Basia", a liaison officer and paramedic who served in "Iwo" Battalion during the Uprising), negative posttraumatic emotions linked with the language of the perpetrators are sharply and directly juxtaposed with the expression of Polish identity by the dying victims. In this fragment of her testimony, she spoke about the time before the Uprising and gave descriptions of street executions by hanging and shooting she had seen, with one of those about to be shot shouting "Poland is not dead yet" (the opening

32 Naliwajek-Mazurek and Spóz, *Okupacyjne losy*, ii:109.

33 An execution on 2 December 1943 at 64 Nowy Świat, close to Świętokrzyska. Police cordons stopped people passing between Ordynacka and Krakowskie Przedmieście. Around fifty inmates of Pawiak jail were shot. Only thirty-four names were made public later in a German announcement (*Bekantmachung*), but more people are known to have been murdered, including the artists Janusz Zoller and Stanisław Haykowski. Bartoszewski, *Warszawski pierścień*, 288–291, 300–301, 316–317. The account of Ludwik Erhardt is quoted in Naliwajek-Mazurek and Spóz, *Okupacyjne losy*, ii:108.

words of the Polish national anthem). She also stated: "I couldn't listen to the German language at all, because I lived through the occupation".[34]

The same witness had a singular auditory memory of the night following the beginning of the Uprising, on 1 August:

> At night I heard something as if a storm was coming: something went, something was coming, a noise, something like thunder. And at some point, from the window – it was warm and all the windows were open – suddenly we heard someone call "Warsaw is free!" It was an amazing sensation.[35]

This memory of sounds resembling the approach of a storm is associated in the witnesses' memories with a premonition of the extreme chaos and omnipresent death during the Uprising, as recounted elsewhere in her account. Unsettling and dangerous as this sound seemed, it was also extraordinary, and it reflected the excitement of the first days of the Uprising and the joyful feeling of liberty that is vividly remembered by most survivors.

Accounts of the traumatic sounds of destruction by the German military arsenal are often combined with descriptions of the effects of such attacks on the inhabitants of Warsaw. Andrzej Kazimierz Olszewski (b. 1931), who after the war became an eminent art historian, remembered the apocalyptic scene of the Warsaw Uprising and the highly vibrating sound of a tank shooting in his direction. Violent images are contrasted in his memory with the fragility of his father's violin, rescued from the ruined house, and the narrative seems to have an impressive, evocative character of core memory, where events are amassed rather than ordered:

> A young man is walking along the corridor with his guts hanging out, shouting: "I'm dying!" A medic runs up to him: "Bronek – [...] – save him!" It was impossible. Our custodian, mister Sierzputowski, a wonderful veteran of the Russian-Japanese War (bearing a striking resemblance to Piłsudski), is saying: "My dears, here at Mokotowskie Field [park], there is a heavy machine gun, and you won't pass through the main gate; it is out of the question. You have to pass through the cellar and run through 34 Polna to 3 Mokotowska, to the yard!" [...] Everybody had his own load, a bit of food, something precious. My father [put] some poetry into his pocket; I kept my father's violin, which was his beloved instrument. When everybody had left, I dropped in for the last package. At that moment, there was a roar so horrible – not a roar, but a sound of high vibration.

---

34 Interviewed in Warsaw on 25 February 2005 by Magdalena Czoch, Sielicka recalled the posttraumatic reaction to the Lithuanian language by a Holocaust survivor, whose testimony is quoted by Robert N. Kraft, "Emotional memory". https://www.1944.pl/archiwum-historii-mowionej/barbara-hanna-sielicka,323.html, accessed 26 September 2017.

35 Interview of 2005, quoted above.

> The tank fired once more; I saw dust and the sky, but I was saved by a sand-filled chest that was standing next to the cellar window. It shattered, but I was saved. But simply that sound of vibration in my ears, and the dust, was awful. Escape, we made it, the injured were lying there, blood, slippery.[36]

Among the various types of warfare sounds described by survivors of the Warsaw Uprising, which together created a terribly loud soundscape, the most specific and most traumatic sound of destruction and death, omnipresent in their accounts, is the sound of a huge German rocket that the Wehrmacht called the *Nebelwerfer*. This name, suggesting smoke or chemical warfare (*Nebel* means fog) served the purpose of disinformation, as the rocket launcher actually carried high explosives, bringing disastrous effects. They were used by the German army against civilian targets in Warsaw in August and September 1944, as well as on several military fronts in Europe. The Germans gave the type used in Warsaw, the sWuR 40, the nickname *Stuka-zu-Fuß*. That referred to the Stuka dive bomber and ground-attack aircraft, which had its own terrifying sound, mentioned by several Warsaw Uprising survivors; this was its stationary variant, "on foot". The residents of Warsaw soon came to recognise it by the loud, terrifying noise it made, but they gave it ironic and dismissive names, "cow" or "wardrobe", reflecting its acoustic characteristics. Horrifying descriptions of the effects of the *Nebelwerfer* on civilian targets are almost pervasive in testimonies from that time.[37] Bogdan Horoszowski (b. 1930, pseudonym "Komar", gunner, "Krybar" group) explained why they aroused such fear:

> Worst of all, I was afraid of the "cows". […] As soon as I heard that the German soldiers were preparing to shoot the cows – it was as if they were moving a wardrobe – then I never went to the shelter, but as soon as the shooting started, I dashed to the cellar, because the shrapnel made you burn, and there was no remedy for it.[38]

---

36 Interview conducted by Joanna Rozwoda on 12 December 2011 and 4 January 2012. Archive of Oral History of the Warsaw Uprising Museum, http://www.1944.pl/archiwum-historii-mowionej/andrzej-kazimierz-olszewski,2805.html, accessed 26 September 2017.

37 This sound led American soldiers to call them "screaming Mimi" or "moaning Mimi". An American Intelligence Report on the Nebelwerfer 41 in an article entitled "Six-Barrel Rocket Weapon" from the November 1943 issue of the *Intelligence Bulletin* signalled: "Its name […] is extremely misleading. In the first place, the Nebelwerfer 41 is not a mortar at all, and, in the second place, it can accommodate both gas-charged and high-explosive projectiles, as well as smoke projectiles", http://www.lonesentry.com/articles/nebelwerfer/index.html, accessed 26 September 2017.

38 Warsaw, 9 August 2006, conversation conducted by Tomek Żylski. https://www.1944.pl/archiwum-historii-mowionej/bogdan-horoszowski,874.html.

Another witness, Wiesław Depczyk (b. 1926, pseudonym "Drohojemski III", army cadet corporal), gave this explanation of the traumatic experience of this rocket's sound:

> On the other side of the Avenue, they set up their hellish roaring "wardrobes" or "cows" […]. The most nerve-wracking thing was the sound of the missiles launching, more than the actual explosions. A roar – and one knew that the missiles would be flying. The firing missiles were hellish: they contained a substance that burned when it exploded a bit and was impossible to extinguish.[39]

Barbara Hanna Sielicka had a particular memory of these and other warfare sounds; her description of the huge mortar and of her survival of a bombardment also features core memory characteristics:

> We suddenly heard that the planes were flying, and the Germans had already begun to bombard more, because there were "cows", there was "Big Bertha". One heard the "cow" as it was being wound up, as we put it, or like a wardrobe was creaking; the creaking was heard and the man was sitting hunched and there was a boom – it was not at us. […] And "Big Bertha" was a huge howitzer, set up somewhere near Warsaw, which fired such huge missiles. Then we heard how this missile flew: a whistling sound was heard as it was flying and after a while a very loud explosion. During this period, the quadrangle between Hoża, Marszałkowska and Skorupki streets was very heavily bombarded.[…] As those bombs dropped, I was blown into the basement through the stairs to the inside. The dust was terrible, you couldn't see a thing. But we lived.[40]

She speaks here of the famous "Big Bertha" (Ger. *Dicke Bertha*), a heavy howitzer used by the German army during the First World War. She was not the only witness to use this name, quite accurately, as years later it was nicknamed "Bertha's Big Brother" by the author of a specialist publication.[41] It was in fact a much more modernised super-heavy self-propelled siege mortar, called "Karl-Gerät", originally conceived to destroy the Maginot Line. Ultimately, the first six mortars (out of seven) were produced only later, from November 1940 to August 1941. It is telling that they had characteristic nicknames, placed on their casing, which ennobled them as a truly German "cultural product" by referring to Nordic mythology, crucial to Nazi ideology and aesthetics. The first two, originally named

39 Warsaw, 7 March 2006, conversation conducted by Iwona Brandt, https://www.1944.pl/archiwum-historii-mowionej/wieslaw-depczyk,659.html, accessed 26 September 2017.

40 Interview of 2005, quoted above.

41 Jentz, *Bertha's Big Brother*.

"Adam" and "Eve", were later renamed "Baldur" and "Wotan",[42] erasing unwanted traces of less noble and powerful figures from the Judaeo-Christian tradition. The remaining four nicknames were taken from Germanic gods: "Thor", "Odin", "Loki" and "Ziu". They were used in Sevastopol and then in Warsaw, where they were transported one by one after initial "successful" trials with "Ziu".[43] One might argue that the ideological violence was thus paralleled and expressed in auditory violence.

A tragic experience of being shot at was narrated by Włodzimierz Kotoński (b. Warsaw 1925; d. Warsaw 2014), who after the war became an eminent composer and composition teacher. During the occupation, he was active in the underground and had lessons with the well-known composer and critic Piotr Rytel. In 1944, he gained his clandestine high school diploma. During the Warsaw Uprising, he was arrested by the German army and deported to the Stutthof concentration camp. He managed to escape from the Death March in which hundreds of Stutthof prisoners perished, but he was subsequently arrested:

> As the front drew nearer, German gendarmes caught me without documents and took me to Kościerzyna. I tried to pass for a civilian worker. They questioned me, asked where I was going. […] They had to return me to Gdańsk, but in the night they received an order to quickly evacuate, and the commander of this small detachment commanded [his soldiers] to "eliminate" [me] and flee. The one who was to kill me turned out to be a Kashubian. He said to me, "don't move, I'll shoot you". He shot into the wall. And they fled.[44]

In the interview, Kotoński did not comment on the sonority of the shooting; one can imagine it from his account, which gives only a bare description of the facts. When asked if he ever referred to his sound memories in his electronic music, which seems to be the most suitable medium through which to express such sonorities, he recalled his characteristic detachment from his war-time experiences and his enthusiasm for the new life. He returned to his war-time experiences only many years later: in conversation, but not in his musical work.

A similar situation was experienced by Halina Grzecznarowska, later to become Władysław Szpilman's wife, who lived in Warsaw after the war, but spent

---

42 "Wotan" had been used as the name of a German tanker during the First World War ("MV Wotan"), and during the Second World War German radio navigation projects were called "Wotan I" (X-Gerät) and "Wotan II" (Y-Gerät).

43 One of the mortars which did not explode hit the Philharmonic Hall and was defused by sappers of the Warsaw Uprising; another one was found recently, in 2012, also in the city centre.

44 Extracts from an interview conducted by the author with Professor Włodzimierz Kotoński, 2014.

the occupation in her native Radom, 100 kilometres from Warsaw. As the front drew nearer, around the turn of 1945, a German army doctor, angry at her mother, who had given some tea to the slave workers who had to dig trenches during the winter for the German-Russian front, shot in the direction of her head on the stairs in front of her house, where she was sweeping the steps. He missed, shooting just above her head. Before he could try again, she was saved by another German soldier, who had been stationed at this villa, where for two months they had lived with a Jewish woman and child in hiding. For several years, the bullet hole remained in the wall.[45] Halina Szpilman also made no mention of the actual sounds of the shooting, simply describing the order of events, reducing her traumatic experience to mere facts.

Music as a tool counteracting traumatic sounds: From anaesthetic to "renormalising"

The majority of witnesses who evoke music during the Warsaw Uprising speak of its importance for boosting the morale of civilians and of those trapped in the city.

One example of such an account is contained in a collective memoir of doctors and women paramedics who – describing the extreme suffering of the wounded and the medical aid that could be offered them – spoke about the positive effects of music in those extreme circumstances. In testimony from 5–18 August 1944, one of the paramedics wrote:

> Soldiers' Day. In the yard of the house at 5/7 Konopczyńskiego Street, a solemn mass is held; in the evening, a concert at the Conservatory at Okólnik Street. The hall is filled to the rafters, although the shooting [outside] continues and the air is full of dust from the falling plaster. Patriotic poems were recited by the actress Karolina Łubieńska,[46] Janina Godlewska sang,[47] her beautiful voice, with its metallic tone, resounding in the dust-filled hall. A man who looked like an artist, perhaps a composer or conductor, taught us all the song "Warsaw children, we're going to fight…"[48] In this hell, there were people trying to bring moments of respite. The Conservatory held concerts, with Mieczysław Fogg and

45 Interview of November 2016.

46 A famous pre-war actress. Her husband, the talented film actor and singer Zbigniew Rakowiecki (1913–1944), was killed during the Uprising.

47 Janina Godlewska, a well-known singer, was the person who saved Władysław Szpilman, and she and her husband, the singer and actor Andrzej Bogucki, received the Yad Vashem title of Righteous Among Nations, on Szpilman's request.

48 The song "Warszawskie dzieci", written by composer Andrzej Panufnik in occupied Warsaw for the underground, to words by Stanisław Ryszard Dobrowolski, one of the best known songs of the Warsaw Uprising.

> Barbara Kostrzewska among the performers.[49] Uprising songs were sung, patriotic poems were recited and Chopin's works were played. The entire hall frequently joined in, singing along to familiar words. On that day, a wounded person was brought to the hospital on Napoleon Square, burned by an incendiary missile – a burning "cow". [...] He was all black, his face covered with carbon crystals [...]. His bed, in Hall One, was screened off, so frightening was the sight of this man in agony [...].[50]

Different reactions to music and its functions in those hellish circumstances were described by another paramedic, Wanda Siedlanowska (pseudonyms "Wanda", "244"), in her account of a hospital arranged in the cellars of an old cafe bar at 25 Podwale in the Old City:

> Sometimes I took a music teacher there. At the "Crooked Lantern", there was a piano on the old stage that we used as a preparatory table to the operating table next to it. At night, sitting on the steps of the stage, leaning against the legs of the piano, paramedics found a moment of rest during the heavy duty hours. The arrival of the professor was great cause for celebration at the "Crooked Lantern". We forgot ourselves in music, a mood of serenity and calm. At the bar, doctor "Bogdan" (a surgeon not known to me personally) and the operating team sipped the wine that was stocked in the cellars. The wounded were lying quietly, forgetting their pain. The nurses took advantage of the short moment of respite. The detonations outside the window did not irritate us; the music made us forget about the horrors of yesterday and today and about the uncertainty of tomorrow. [...]
>
> I remember that date exactly. It was August 24th. The music teacher came and started his concert, so we had a short rest. I was sitting in my corner with "Tadeusz" (Tadeusz Tan, a corporal army cadet from "Aniela" Company), whose right-arm *vulnus sclopetaria* [actually *sclopetarium*, gunshot wound] refused to heal, when suddenly the music stopped in mid-bar. Two wounded men were brought in. We all jumped up. [...] The doctor raises his head and cries to the nurses gathered around the table: "More light, operation!" [...] The wounded man summons me with a motion of his hand: "Hold me by the head", he asks. And then, when I lean over him, he asks, "Let him keep playing". The doctor agrees; it won't disturb him. He is already concentrating, moving his finger slowly down the patient's thigh, as if marking where he is about to run the scalpel. [...] It lasts an eternity, or perhaps just a moment. You can hear the bright, joyous voice of the doctor: "Nurse, dressing!" He holds a small short bullet in his fingers, probably of a machine gun. He is very pleased with himself: "I wasn't even a millimetre out", he says. And turning to the patient: "Did it hurt?" "What?" The fingers clenched on my shoulder slowly relax, his facial muscles are sagging: "It's fine now", he says with an effort, then to the doctor, the teacher and all of us: "Thank you". I have tears in my eyes. This is music that tears me

49 Stars of the pre-war and post-war stage.

50 Łubek-Luboradzka, Mieczkowska and Gleb, *Sanitariuszki*, 46–7.

apart. Zdzisław [the other patient] died half an hour later, without regaining consciousness. Over his grave, late at night, we sang "Rota" with strong voices.[51]

This testimony speaks of several positive functions of music in traumatic circumstances: its anaesthetic power and healing properties, and its ability to move and connect with one's feelings (with the "side effect" of losing composure), to raise morale and to commemorate a fallen colleague.[52]

Singing as a method to counteract the traumatic sounds of warfare

Another witness who took part in the fighting, Wanda Traczyk-Stawska (b. Warsaw, 1927), spoke about the special role of singing as a loud sound effect used to drown out the sounds of German bombardments. The described event took place on 6 September 1944:

> There was a horrible bombardment. Boys and girls not involved in the fighting were singing as loud as they could, so that the bombardment would not be heard. It was then that I heard for the first time *Gaudeamus igitur*.[53]

One child survivor of the Holocaust, Halina, recalled the significance of music during the last months of 1944, outside Warsaw, which constituted an effort to return to a form of more normal life and to recreate the identity of the homeless Varsovians:

> After the surrender [of the Uprising], when we were being taken to Pruszków, we managed to escape. Some people in Szczęśliwice helped us escape. They took us in for a night and stole our things. With some friendly neighbours, we went to Milanówek, where an acquaintance of theirs took us to his villa. The villa was full of people, but there was a place to sleep on the kitchen floor. The house was very welcoming, and the cultural life was flourishing. Professor Rączkowski (a famous organist) played piano, several children – already teenagers – sang patriotic songs and chants, and poetry was recited. I even attended concerts held in Milanówek – in private homes, of course.[54]

---

51 Ibidem, 28–30. "Rota" [The oath], with words (1908) by Maria Konopnicka and music (1910) by Feliks Nowowiejski, is treated as sort of anthem of the Polish fight for independence against the Germans.

52 On the positive functions of music during the Holocaust, cf. e.g. Eisen, *Children and Play*; also Fisher and Gilboa, "The roles of music".

53 Interview conducted in Warsaw, on 8 July 2009, by Mariusz Kudła, https://www.1944.pl/archiwum-historii-mowionej/wanda-traczyk-stawska,60.html, accessed 26 September 2017.

54 Meloch and Szostkiewicz (eds.), *Dzieci*, 154.

Imagined music: Musical memory as a survival technique

The writer and journalist Wacław Gluth-Nowowiejski (b. Warsaw, 1926) took part in the Warsaw Uprising as the commanding officer of one of squads of the "Żmija" group in Żoliborz (a northern district of Warsaw). Two of his three brothers were shot in July 1944 in a public execution.[55] The only one of the four brothers to survive, Wacław, was first wounded on 14 September, and he was treated by paramedics sent there by the commandants. There, at 12 Rajszewska Street, German soldiers massacred the wounded with hand grenades and then killed them one by one with machine guns. They also shot all the paramedics, although the girls had visible Red Cross bands on their arms. Wacław Gluth-Nowowiejski was shot in the head and was lying among his killed friends. The Germans thought he was dead. In spite of the lack of food and his wounds, he survived until mid-November 1944, hiding in the cellar of a house at what was then 6 Warszawska Street. Lying in sort of a bed, he was considered dead by all the German or ROA soldiers who found him, before being rescued by a Polish woman.[56] While lying half-conscious in that cellar, he imagined music:

> First of all, I lost track of time, I didn't know […] how I had got there, and I didn't even think about it. I stopped thinking. I knew it was cold, that there was a gale and it was snowing outside […]. Yet I was lying there, and there were moments of complete sobriety. I would say that it is quite a strange thing, and today for me personally it is extraordinary – because I've always loved music, so all of a sudden I began to recall different classical works, as I'd attended a music school during the occupation. It was something completely out of this world.[57]

---

55 One of the four brothers, Janusz, was a pianist, who studied before the war with Bolesław Woytowicz at the Warsaw Conservatory and during the war worked with his brother as a cloakroom attendant at the famous musical cafe run by Woytowicz, and they both belonged to the underground. The third brother, Jerzy (pseudonym "Pigi"), perished as a soldier of the Uprising (a corporal army cadet) on 9 September at the Church of the Sisters of the Visitation.

56 After the war, in November 1948, he was arrested for being a member of the Home Army. Imprisoned, he endured cruel interrogations, before being released in 1953.

57 Wacław Gluth-Nowowiejski, in an interview of 20 February 2007, conducted by Małgorzata Brama, for the Archive of Oral History of the Warsaw Uprising, http://www.1944.pl/archiwum-historii-mowionej/waclaw-gluth-nowowiejski,1054.html, accessed 26 September 2017. He has also narrated his experiences in several books: *Śmierć poczeka* [Death awaits], *Rzeczpospolita gruzów* [The republic of ruins], *Nie umieraj do jutra* [Don't die till tomorrow] and *Stolica jaskiń: z pamięci warszawskiego Robinsona* [Capital of the caves: from the recollections of a Warsaw Robinson Crusoe].

In another conversation,[58] he explained that the sounds he then imagined were Schumann's piano music and some of the opera music that his brother Janusz liked to sing at home. There was no Chopin in his musical memory at that time, as if the ban on playing Chopin was present somewhere in his unconscious. This may seem paradoxical, especially since the only time that ban was lifted for Woytowicz's café was for an emotional and memorable concert attended by Wacław Gluth-Nowowiejski, when almost everybody was crying.[59] This psychological mechanism of imagination, resembling the tactic used by Władysław Szpilman in hiding, is a fascinating example of survival strategies. As Szpilman wrote about the same time as Gluth-Nowowiejski (early November 1944), where he was hiding in the building at Niepodległości Avenue:

> I lay motionless all day long to conserve what little strength I had left putting out my hand only once, around midday, to fortify myself with a rusk and a mug of water sparingly portioned out. From early in the morning until I took this meal, as I lay here with my eyes closed, I went over in my mind all the compositions I had ever played, bar by bar. Later, this mental refresher course turned out to have been useful: when I went back to work I still knew my repertory and had almost all of it in my head, as if I had been practising all through the war.[60]

Szpilman's specific experience of remaining in hiding and guessing what was outside from the sounds he could hear lends his testimony a vivid soundscape character, where military changes are reflected in his perception of the decreasing volume of Soviet military activity. Between June and July, he remembered: "Soviet raids on Warsaw came more and more frequently; I could see the fireworks display from my window. There was a growling noise in the east scarcely audible at first, then growing stronger and stronger: Soviet artillery". On 1 August 1944, it changed: "the thunder of the Soviet artillery, so close a few nights earlier, was now clearly moving away from the city and becoming weaker".[61]

Traumatic sound as a trigger to creativity

A special account of traumatic sounds of particular importance in the survivor's subsequent biography was given by Halina Paszkowska-Turska (1927–2017), who

58 A conversation with the author of the chapter, in 2017.

59 This account was given by Bolesław Woytowicz in *W okupowanej Warszawie* [In occupied Warsaw], see annotated version in Markowska and Naliwajek-Mazurek, *Okupacyjne losy*, i.

60 Szpilman, *The Pianist*, 168.

61 Ibid., 149 and 151.

became an outstanding film sound engineer in the post-war era.[62] Imprisoned in the Warsaw Ghetto during the occupation, she then survived in hiding under a false name, which she retained for the rest of her life. During the Warsaw Uprising, in 1944, she belonged to the "Rafałki" storm platoon of the "Konrad" group in the Powiśle riverside district. In an interview recorded at her home in November 2016, she gave the following account, beginning with a description of the post-war landscape:

> Yes, one walked among the ruins. People immediately set about reconstructing the city. It was incredible. Even today, I cannot imagine such a city as Warsaw being raised from such ruins! […]
>
> The armoured train that passed over the nearby bridge created terrible havoc, shooting from machine guns. We received a message that there was something going on out there at the [National] Museum, next to Smolna Street. They had apparently arrested all the men, marched them into the Museum and shot them. We didn't have much idea where the Germans were and where something needed to be done to rid the streets of gunfire […].
>
> Our commander sent a group of five or so: four or five boys and myself. At night, we went through the streets next to St Elizabeth's. We were supposed to go along Smolna Street to enter the buildings vis-à-vis the Museum. Because that street was under fire, we couldn't just march there all together, but had to simply slip through. Each of us separately rushed out and burst through the gate. I was there too, and I had to do it.
>
> When I passed through the gate, it was filled with debris and glass. Every step sounded like a shot. Otherwise, there was complete silence. We wanted to enter those houses to see what was happening out there, who was there, whether there were any Germans.
>
> At some point, while moving across this broken glass and rubble, I saw an open door vis-à-vis with a view of the Museum! […]. At one point, I was preparing to go when I saw smoke, as if from a cigarette, in the passage. It could only have been a German, so I stopped. I froze completely and stopped breathing. I stood like that for a couple of minutes, and of course everything was running through my head. I was wearing a helmet, strap and everything, so there was no doubt who I was.

---

62 She worked as a sound engineer on about twenty drama films, including Roman Polański's famous *Knife in the Water* (1962), nominated for Best Foreign Language Film at the Academy Awards in 1963, Aleksander Ford's *The First Day of Freedom* (1964), an entry in the 1965 Cannes Film Festival, and more than two hundred documentary films created by such outstanding directors as Kazimierz Karabasz (incl. *The Musicians*, 1960) and Marcel Łoziński *Zderzenie czołowe* [Head-on collision] (1975), *Wszystko może się przytrafić* [Anything can happen] (1995). In 2010, she received the Prize of the Polish Film Association. She was the wife of historian Marian Turski, a survivor of Łódź Ghetto and Auschwitz, founder of the eminent Polin Museum in Warsaw, an editor of the weekly *Polityka*.

> I stood stock still, so as not to awake this devil. It was a horrible moment, because I knew it might be the last moment of my life. However, as it took quite a long time, I came to the conclusion that it was not a cigarette, but ashes, so I began to breathe a little.
>
> So that is the dramatic sound which stayed in my head. I was thinking that if I survived, I would like to render the drama of those steps, the drama of that sound [...]. I was thinking about telling it to someone one day.
>
> And then it turned out that I went into sound engineering and film and spent the rest of my life doing nothing but sounds![63]

One can argue that it was Paszkowska-Turska's creativity that allowed her to mould her traumatic sonic memory into a passion for creating complex artistic narratives in sound and to "restore integration of her self". Her highly intuitive way of working with unsophisticated recording equipment brought striking auditory results, which played an important, often crucial, role in the narrative of films. Such was the case with the famous scene in Polański's *Knife in the Water* when she succeeded, during the post-synchronisation process, in merging the voices of Andrzej (Leon Niemczyk) and Jolanta Umecka (Krystyna, his wife) with the voice dubbed by actress Anna Ciepielewska, the *mormorando* song composed by Krzysztof Komeda (sung by Halina Paszkowska-Turska herself, who had excellent musical memory) and the calling of the young man (Zygmunt Malanowicz), who has supposedly drowned but is actually hiding in the water to scare them. In several other films, she invented dramatic effects, guided by her training and imagination. The eminent documentary film-maker Marcel Łoziński spoke of her persistence in recording effects and of her mastery, as with the crunching of broken glass in *Zderzenie czołowe* [Head-on collision] (1975). This climactic moment in the film evokes a traumatic moment from the life of the main character, a train driver who caused a railway accident six months before retiring. Łoziński described this sound effect as "realistic, but at the same time acquiring a higher artistic dimension, which is magical".[64] This artistic "sound itinerary" might be interpreted as an effective strategy for integrating traumatic memory into the difficult reconstruction of "a sense of the culture in which the trauma occurred", achieving "recovery within an embedded cultural framework".[65]

Another trait of crucial importance for posttraumatic reintegration of the self is Halina Paszkowska-Turska's choice regarding her surname. Although she could have returned to her original name Penczyna after the war, she decided to retain

---

63 Interview conducted by the author, November 2016.

64 Speech given during Halina Paszkowska-Turska's anniversary celebration in June 2017.

65 Wilson, *An Overview*, 78–9.

the name Paszkowska, which was offered to her by an unknown young woman, so that she could survive under a false identity as an "Aryan". The young woman probably died during the occupation, because one day she disappeared from the tram route where they used to meet, so Paszkowska-Turska decided to keep the name as a sign of memory and gratitude, even though her original name did not sound Jewish and would not have betrayed her Jewish origins even after the war.[66] In this symbolic way, she also created a "new self" that merged her identities and the desire to commemorate her helper and led to what Fisher and Gilboa term identity re-formation after the Holocaust.[67]

The sounds of occupied Warsaw versus the sounds of the ruined city

It is evident that the soundscape of the silent, ruined Warsaw after the war vividly contrasts in witnesses' memories both with the acoustic image of the noisy city, still partially "alive" before August 1944, and with the horrifying soundscape dominated by German artillery and bombing during the Warsaw Uprising. Musicologist Władysław Malinowski, who was eight when the war began and thirteen when it ended, remembers post-war Warsaw as a mountain-like landscape filled with snow and silence. That changed in the spring, when surrealistic images emerged, like in paintings by Chagall. The first music he heard was from LPs that he bought in a store, opened already in 1945.[68]

Yet some of the remembered sounds from the first days after 17 January still possess the alarming quality linked with a life-threatening meaning. Among her experiences on returning to Warsaw in 1945, the composer's wife, harpist Jadwiga Kotońska, described the specific sounds of what was probably a Wehrmacht soldier's rifle:

> We arrived in the western part of Warsaw on 17 January. It was totally empty. Suddenly we saw a miserable-looking German walking along and heard [the sounds of a rifle hitting the railings]. And my mother said "listen, let's run away; he might pick up the rifle and point it at us". We went back and returned the next day.[69]

---

66 Information provided during an anniversary celebration, held on 22 June 2017 at the Kultura Cinema in Warsaw, with speeches given by Marcel Łoziński, among others.

67 Cf. Fisher and Gilboa, "The roles of music".

68 Ibid.

69 The above-quoted interview with the composer and his wife in 2014.

The actor Witold Sadowy (b. Warsaw, 1920) described the contrast between the noisy occupied Warsaw, full of the sounds of trams and rickshaws, with the ruined city immersed in darkness and silence that he found on returning to Warsaw on 19 January 1945.

He vividly remembers concerts and theatre performances, which began to take place in the ruined city. At a small theatre on Marszałkowska Street, an opera was staged in December 1945: "it was such an incredible event that the foreign press wrote about it – operas being performed in the ruins of Warsaw".[70]

Traumatic sound memory as a component of creative transformation and a vehicle of self-integration

In this short study, only a limited selection of examples was quoted, and only a few questions concerning the relationship between memory and trauma could be signalled. Interviews with survivors demonstrate that auditory memory of bombings and shootings – including the most direct and most traumatic experience of being shot at – usually accompanies or replaces visual experiences. Certain sounds linked with intense emotions such as imminent danger are remembered and described by some witnesses in an acute way. Survivors describe the sounds of bombing and killings, music present in several places during the Uprising and stories of music after the escape from Pruszków transit camp.

Although incomplete, this overview of sound memories gives some idea of the effectiveness of the sound violence inflicted by German warfare in Warsaw and how it was intentionally moulded as the ideological occupation of symbolic and physical spaces.

In the case of survivors sensitive to auditory experiences or stimuli, sound memory may become not only significant material integrated into a personal narrative but also a crucial component of intrapersonal soundscape topography, with chronologically ordered events organising its timeline.

Memories composed of sounds, among other phenomenal experiences characteristic of a highly traumatic environment, have the vivid, immediate character of core memory. What is more, in some instances they acquire a special, creative potential, as Jenny Edkins suggests:

> Events of the sort we call traumatic are overwhelming but they are also a revelation. […] They question our settled assumptions about who we might be as humans and what we

70 Ibid.

> might be capable of. Those who survive often feel compelled to bear witness to these discoveries.[71]

This paradoxical statement finds corroboration in the testimonies and post-war biographies of survivors, especially Halina Paszkowska-Turska. In several cases, their traumatic experiences remembered through sound memories formed part of their highly creative healing processes. A sense of posttraumatic discontinuity leading to an "old self" being replaced by a "modified/different/new self" can be observed in most survivors' accounts.

Furthermore, some witnesses' descriptions and explanations of music's functions highlight its trauma-reducing properties, in the form of improvised concerts, as in accounts of Warsaw Uprising paramedics, or in imaginary form, as in cases of people immobilised and cut off from life, like Szpilman and Gluth-Nowowiejski. The issue of the roles of music in traumatic conditions is extremely complex. Its negative roles, mainly consisting in the abuse of music, have been analysed, for example, with regard to the death camps,[72] whereas several positive functions for the survival of camp prisoners often appear in their testimonies, when music became a strategy of survival, a means of strengthening resilience.[73] The above-quoted accounts confirm that music may function as a "normalising" factor in traumatic circumstances. However, due to the heterogeneous character and origin of these sources, it cannot be demonstrated that passages describing such functions of music are structured in a different way than accounts of traumatic sound experiences or nontraumatic memories.

Given the limited scope of the interviews and their historically-oriented character, it is impossible to ascertain the extent to which verbalised memory was useful in healing processes and whether such memories may act as cues for posttraumatic remembering. It seems plausible, however, that the ability to transform a traumatic sound event into a potent factor stimulating creativity might serve as a motivational agent to overcome the trauma and build a new identity, integrating positive and negative experiences into a creative whole. The testimonies quoted here demonstrate the cases of individuals who succeeded in transforming their victim identity into one of a survivor – a new creative self of performer, artist or writer.

---

71 Edkins, *Trauma*, 5.

72 See the literature quoted by Fisher and Gilboa, "The roles of music", including Naliwajek-Mazurek, "Music and torture", 31–50. A more general review of these topics is contained in Naliwajek-Mazurek, "The functions of music", 83–103.

73 Cf. Eisen, *Children and Play*, Fisher and Gilboa, "The roles of music".

Bibliography

Diagnostic and Statistical Manual of Mental Disorders (Washington, DC: American Psychiatric Association, 2013).

Bartoszewski, Władysław, *Warszawski pierścień śmierci 1939–1944* [The Warsaw ring of death 1939–1944] (Warsaw: Interpress, 1970).

Berger, Roni, *Stress, Trauma, and Posttraumatic Growth Social Context, Environment, and Identities* (New York: Taylor and Francis, 2015).

Birdsall, Carolyn, *Nazi Soundscapes. Sound, Technology and Urban Space in Germany, 1933–1945* (Amsterdam: Amsterdam University Press 2012).

Dybowski, Stanisław, *Słownik pianistów polskich* [Dictionary of Polish pianists] (Warsaw: Selene, 2003).

Edkins, Jenny, *Trauma and the Memory of Politics* (Cambridge: Cambridge University Press, 2003).

Eisen, George, *Children and Play in the Holocaust. Games among the Shadows* (Amherst: University of Massachusetts Press, 1988).

Fisher, Atarah, and Gilboa, Avi, "The roles of music amongst musician Holocaust survivors before, during, and after the Holocaust", *Psychology of Music*, 44/6 (2016), 1221–1239.

Good, Byron J., and Hinton, Devon E. (eds.), *Culture and PTSD: Trauma in Global and Historical Perspective* (Philadelphia: University of Pennsylvania Press, 2016).

Jentz, Thomas, *Bertha's Big Brother: Karl-Geraet (60 cm & 54 cm)* (Boyd, MD: Panzer Tracts, 2001).

Kraft, Robert N., "Emotional memory in survivors of the Holocaust: A qualitative study of oral testimony", in Daniel Reisberg and Paula Hertel (eds), *Memory and Emotion* (Oxford: Oxford University Press 2004).

Laub, Dori, "From speechlessness to narrative: The cases of Holocaust historians and of psychiatrically hospitalized survivors", *Literature and Medicine*, 24/2 (2005), 253–65.

Łubek-Luboradzka, Teresa, Mieczkowska Krystyna, and Gleb, Barbara, *Sanitariuszki Batalionów AK Gustaw i Harnaś w szpitalach powstańczych. Wybór wspomnień sierpień–wrzesień 1944* [Female paramedics of the Home Army battalions "Gustaw" and "Harnaś" in the hospitals of the Uprising. A selection of memoirs August-September 1944] (Warsaw: nn, 1995).

Markowska, Elżbieta, and Naliwajek-Mazurek, Katarzyna, *Okupacyjne losy muzyków, Warszawa 1939–1945* [Musicians in occupied Warsaw 1939–1945], i (Warsaw: Towarzystwo im. Witolda Lutosławskiego, 2014).

Meloch, Katarzyna, and Szostkiewicz, Haline (eds.), *Dzieci Holocaustu mówią...* [Children of the Holocaust talk...], iii (Warsaw: Biblioteka Midrasza, Stowarzyszenie Dzieci Holocaustu w Polsce, 2008).

Mościcki, Tomasz, *Teatry Warszawy 1944–1945. Kronika* [Warsaw theatres 1944–1945: a chronicle] (Warsaw: Fundacja Historia i Kultura, 2012).

Naliwajek-Mazurek, Katarzyna, "Music and torture in Nazi sites of persecution and genocide in occupied Poland 1939–1945", *The World of Music* (new series), 2/1 (2013), 31–50.

– "The functions of music within the Nazi system of genocide in occupied Poland", in Wojciech Klimczyk and Agata Świerzowska (eds), *Music and Genocide*, Studies in Social Sciences, Philosophy and History of Ideas, 9 (Frankfurt am Main: Peter Lang, 2015).

–, and Spóz, Andrzej, *Okupacyjne losy muzyków, Warszawa 1939–1945* [Musicians in occupied Warsaw 1939–1945], ii (Warsaw: Towarzystwo im. Witolda Lutosławskiego. 2015).

Raphael, Beverley, and Dobson, Matthew, "Acute posttraumatic interventions", in John P. Wilson, Matthew J. Friedman and Jacob D. Lindy (eds), *Treating Psychological Trauma and PTSD* (New York: Guilford, 2001).

Snyder, Timothy, *Bloodlands: Europe between Hitler and Stalin*, 2nd edn (New York: Basic Books, 2012).

Szpilman, Władysław, *The Pianist. The Extraordinary True Story of One Man's Survival in Warsaw, 1939–1945*, tr. Anthea Bell (New York: Picador, 1999).

Tulving, Endel, and Craik, Fergus I. M., *The Oxford Handbook of Memory* (Oxford: Oxford University Press, 2000).

Van der Kolk, B. A., and Fisler, R., "Dissociation and the fragmentary nature of traumatic memories: Overview and exploratory study", *Journal of Traumatic Stress*, 8/4 (1995), 505–25.

Whitehead, Anne, *Memory* (London: Routledge, 2010).

Wilson, John P., "An overview of clinical considerations and principles in the treatment of PTSD", in John P. Wilson, Matthew J. Friedman and Jacob D. Lindy (eds), *Treating Psychological Trauma and PTSD* (New York: Guilford, 2001).

Young, Allan, *The Harmony of Illusions. Inventing Post-Traumatic Stress Disorder* (Princeton, NJ: Princeton University Press, 1997).

Jadwiga Zimpel

Institute of Cultural Studies, Adam Mickiewicz University

In search of lost sounds: Miron Białoszewski's "Stare życie" and post-war silence

Abstract: This paper is devoted to reflection on post-war silence, which I propose to understand as an "earcon" of absence. An attempt to reconstruct the meanings attached to post-war silence by earwitnesses is undertaken through a reading of the series of short stories by Miron Białoszewski entitled "Stare Życie" [Old Life], in which the auditory experience of silence is interpretatively related to the notions of melancholy, mourning and trauma. The last section of the article discusses the idea of literature as sound memory and reflects on the possible advantages of the adoption of a hauntological perspective in studies on historical soundscapes.

Keywords: silence, earcon, literature, sound memory, hauntology

In this paper, I focus on silence, considered both as an experiential feature of the post-war urban soundscape and as a multi-dimensional category that might be used to describe post-war auditory experience. I excavate the meanings of this contradictory phenomenon through a reading of the collection of short stories entitled "Stare Życie" [Old Life] by the Polish poet and novelist Miron Białoszewski. Reflecting on Białoszewski's decision to ground the narration of "Old Life" within the time-frame of the pre-war soundscape, I seek to problematise that gesture as a paradoxical response to the experience of post-war silence. One of the themes of the oeuvre of Miron Białoszewski, who was a member of the Generation of Columbuses, is the experience of the war. Records of that experience can be found in works such as *A Memoir of the Warsaw Uprising* and "Old Life", in which Białoszewski reconstructs not so much the war itself, but the world it destroyed. Literary historians ascribe Białoszewski's work to the current of linguistic poetry, although it should be emphasised that the original style of his writing evades unambiguous classification. Białoszewski accentuates the importance of the spoken word as the basis of his poetry and prose, as expressed, for example, by the use of a tape recorder as one of his writing tools. The rhythm and melody of language are highlighted by the writer through surprising word formations and references to the conventionalised uses of language that structure the everyday life of certain cultural communities, such as urban residents, passers-by or worshippers.

The methodological approach to the notion of historical soundscape that I adopt in this paper relates to Mark M. Smith's conception of "heard worlds".[1] Thus it is based on a sound studies approach to historical auditory experience, which contrasts with the approaches of acoustic ecologists. Smith understands "heard worlds" as phenomena which require in-depth contextualised investigation not restricted to the acoustic measurements that characterise Schafer's school.[2] According to Smith, in order to cultivate culturally informed sound history, one needs to investigate the "social and historical implications of heard worlds" and to adopt "subjective or cognitive understandings of sound".[3] Drawing on Smith's concept, one notes that the specificity of the auditory experience of a certain moment in history is closely related to the cultural and social positioning of the listener and to his or her cultural identity. The heard worlds are thus deeply rooted in the network of relationships and sets of circumstances within which and through which the subject listens. Taking the mutual dependence of ear and culture into consideration, we can also state that the quality of subjective auditory experience and the action that is taken by the subject in response to it is mediated by what symbolic interactionists would call the "definition of the situation", which is an outcome of participation in a certain cultural universe and helps subjects to communicate with others. As a mediation of the auditory experience, the definition of the situation might also be considered as a symbolic basis on which the subject gives meaning to what he or she has experientially encountered.

Viewed from the perspective of sound studies, the historical auditory experience constitutes a multi-dimensional object of study. Its reconstruction involves the consideration of differences among historical listeners which might affect the way they define the geographical and historical context in which they hear sound. Thus in studying phenomena such as post-war silence, one needs to take into account the specific subjective cultural context in which it is experienced, including the spatial, social and axiological positioning of the earwitness. Such a socio-cultural construction of an historical soundscape is characterised by interpretative openness. If a soundscape is a "composition", as R. Murray Schafer once noted,[4] it is a composition heard through various cultural filters. From such a perspective, an historical soundscape might be also considered as an "open work", to use the well-known formulation of Umberto Eco.[5] As such, it is an object of

1 Smith, "Listening", 137.
2 Ibid., 138.
3 Ibid.
4 Schafer, Introduction, 5.
5 Eco, *The Open Work*.

multiple interpretations, including in the form of translation into another media, as in the case of literature.

My methodological assumption regarding literary practice and its relation to auditory experience follows the conception of Philipp Schweighauser. He argues that literature not only represents the soundscape of its time, but it is also a means of its construction. In his research into the modern and postmodern novel, Schweighauser analyses selected literary works as specific types of sound sources. "This was most evident – he writes – in modernist texts […]. Not only these texts but also earlier, naturalist texts […] and postmodernist texts […] go beyond an attempt to represent the soundscape of their time and place. In their rhythmic structure; their jarring juxtapositions of different media, genres, and styles; and their textual dislocations and fragmentations, these texts themselves become sounding objects".[6]

It is perhaps important to add that the discussion with the concept of literary acoustics as representations of a soundscape has its roots in reflection on modern urban experience. Elżbieta Rybicka writes that this experience was mostly perceptual, and in order to articulate its nature in literature, writers needed to invent a new type of artistic expression.[7] She proposes to consider their linguistic experiments as "responses" to urban stimuli.[8] Thus literature functions not as a separate cultural space, but rather as another layer of an "imagined city" – to refer to the concept of James Donald.[9] From such a point of view, it is obvious that for a writer to grasp his or her auditory experience of a certain moment the question of creation or working through is more important than the question of representation.

Silence as the earcon of absence

Post-war Warsaw, like other destroyed cities, functions in the collective imagination as an icon of absence, while silence might be regarded as its "earcon".[10] Barry Blesser and Linda-Ruth Salter define the earcon as "a sonic event that contains special symbolic meaning not present in the sound wave". They also state that earcons operate as linkages between "the here" and "the there".[11] Blesser and Salter consider the earcon with regard to the human-architecture relationship. They use this category as a tool that brings out the connection between the symbolic

6 Schweighauser, "Literary Acoustics", 476.
7 Rybicka, *Modernizowanie*, 100, 102.
8 Ibid., 20.
9 Donald, *Imagining*.
10 Blesser and Salter, "Ancient Acoustic Spaces", 187.
11 Ibid.

meaning of sound, cultural practices and their socio-spatial context. Hence what determines the symbolic meaning of the earcon, following the reasoning of Blesser and Salter, is the specific cultural practice that engages a given community in a particular space. Blesser and Salter note that earcons differ in terms of the range of their symbolising functions. To them, both the sound of a bell summoning household members to a family meal and the reverberation characteristic of buildings of sacred architecture might be considered as earcons. In both cases, the symbolic meaning of an earcon is legible only to members of a particular community engaging in specific practices or events. Thus the proposed approach to silence as an earcon of absence employs not the notion of abstract silence, but rather a historically and culturally contextualised silence. Thus I am interested in silence that was experienced by inhabitants of post-war Warsaw who remembered how the city sounded before the outbreak of war.

One might notice that the earconic relationship between silence and absence is not self-explanatory. It is rather interpretatively evoked by various types of cultural texts, which deal with post-war trauma. In his analysis of architectonic expression, Witold Szymański asserts that the meaning of silence as absence, void or emptiness is conveyed by commemorative architecture.[12] To Szymański, exemplary in this regard is the Jewish Museum in Berlin designed by Daniel Libeskind.[13] Besides architecture, the relationship between silence and absence is also present in other texts of Western culture, for example in poetry.[14] One also notes that the idea of absence is deeply related to the idea of death. Death-related connotations of silence are accentuated by Mark Slouka, among others. He explains his ambivalent attitude toward the discussed phenomenon in the following words: "the grave, the scythe, the frozen clock, all the piled symbols of death, reinforce an essential truth, a primal fear: beneath the sloping hood, death is voiceless".[15] According to Slouka, the voicelessness of death explains why in some contexts people often fear silence. He argues that this fearful or even resistant attitude toward silence might be considered as a struggle against nothingness.[16]

But does silence really exist? "There is no such thing as empty space or an empty time. There is always something to see, something to hear. In fact, try as we may to make a silence, we cannot".[17] These well-known words of John Cage

---

12 Szymański, "Cisza w architekturze", 71.

13 Ibid.

14 Jasionowicz, *Pustka.*

15 Slouka, "Listening for Silence", 42

16 Ibid.

17 Cage, "Experimental music", 8.

contradict the idea of absolute silence, as well as absolute absence. So can we be justified in thinking about a post-war city as silent, still, quiet?

One might argue that the idea of post-war silence is a speculative construct, because everyday life in ruined cities did not stop, and it probably generated multiple forms of daily noise. However, I propose that the place of silence as an experiential feature of the post-war soundscape might be asserted while taking into consideration the sounds that preceded it. In more general terms, such an observation is also shared by Robert Losiak, who states that the experience of silence is connected with the differences in dynamics between consecutive sound phenomena.[18] Similarly, Katarzyna Szalewska has shown, referring to Carolyn Birdsall, that the experience of post-war silence had a relational character.[19] This means that it was linked with the experience of war noise. Birdsall writes: "While descriptions of the end of the war emphasize silence and absence, the wartime soundscape was profoundly dominated by the three main sounds of sirens, planes and bomb attacks".[20] Considering this relational quality of silence, we can suppose that due to the contrast with the often disorganised noise of war, post-war manifestations of silence must have been acutely experienced.

In order to understand post-war auditory experience, it would appear important to reflect on the psychological and socio-cultural effects of the contrast between the auditory qualities of wartime and peacetime. One might think about the juxtaposed realities of noise and silence in terms of a structural opposition between presence and absence, life and death, chaos and order. The cessation of the noises generated by military action did not necessarily bring the anticipated relief. For example, in descriptions of the end of the war cited by Birdsall, silence is often connected with the images of a ruined landscape and destroyed environment. Such an observation reveals another layer to the meaning of post-war silence: feelings of regret, grief or even dismay, which are the outcome of the absence of pre-war communities.

Thus silence as an earcon of absence is a socio-cultural phenomenon, not merely an acoustic one. As such, it might be used as a descriptive category of the lack of expected, well-known or even beloved sounds, not sound in general. So the symbolic meanings attached to post-war silence by earwitnesses will differ according to the subjective definitions of the nature of experienced loss.

---

18 Losiak, "Cisza", 60.

19 Szalewska, "Pejzaż", 36.

20 Birdsall, *Nazi Soundscape*, 129.

Silent about silence

Published in 1980, the series "Old Life" is part of the book *Rozkurz* [Wastage].[21] It consists of three short stories: "Leszno 99", "Nanka" and "Opowiadanka z niepokojem" [Disturbing tales]. An especially vivid depiction of the heard world of pre-war Warsaw can be found in the first of the "Old Life" stories: "Leszno 99". Using the strategy of detailed, precise, unbiased description, Białoszewski recreates the ambience, spatial ordering, blood relations and cultural characteristics of the world of his childhood. Some might observe similarities to Walter Benjamin's *Berlin Childhood around 1900*. Skilfully written passages create an illusion of autobiographical insight. Small, seemingly meaningless and obvious elements of everyday life are transformed into something new and imbued with magical meaning. The author's ordinary usage of language, with its idiosyncrasies, preserves the memory of people who have long gone. Description involving different senses, with an ear for the richness of everyday sounds, makes Białoszewski's narrative a complete cultural space, in which what he calls old life takes place over and over again, remaining untouched regardless of external circumstances. The reader is invited to experience the interrelated qualities of safety and dynamics that constitute the presented world. Descriptions of a linen press, with its specific "du-du-du-du" sound,[22] a gramophone playing somewhere in the tenant house with its "wu-wu-wu",[23] the tone of a neighbour's laughter,[24] the loud clicks of closing purses,[25] the barks of a dog and the whooshing of the wind all constitute the soundscape of a safe, protected place with clearly defined boundaries – a utopia of sorts. But is it really?

As a result of the war, the world presented in "Leszno 99", with its people, ambiences and topographies, was destroyed. After the liquidation of the Warsaw Ghetto, where Leszno Street was located, and as a consequence of the post-war rebuilding, the topography of this part of Warsaw changed. In the visible fabric of contemporary Warsaw, there is no trace of the place of Białoszewski's childhood. However, memories of it have been retained in writing.

"Old Life" thematises silence by means of its relational quality, which, as already mentioned, is based on its mutual relationship with sound. Białoszewski then paradoxically stages the silence by setting the narrative timeframe outside the post-war reality. Such a gesture might be considered as partly escapist, because it

21 Białoszewski, *Rozkurz.*

22 Białoszewski, "Stare Życie", 6.

23 Ibid., 9.

24 Ibid.

25 Ibid., 7.

is based on a return to the magical soundscape of the writer's childhood. Thus the experience of loss that concerns both the lost childhood and the tragedy of the war is questioned here by the omission of post-war silence as a theme in the analysed stories and its replacement with the voices of the pre-war period. Such a gesture might be interpreted as a subjective way of dealing with trauma. From such a point of view, it might be noted that "Old Life" employs the sonorous qualities of literary expression not to represent auditory experience of the pre-war period, but rather to alter the post-war soundscape. The sonorities of language and other means of literary expression are utilised by the writer to rebuild the ruined Warsaw in words. Thus, through his writing, Białoszewski is searching for lost sounds with which to fill the post-war silence. Such a strategy makes his attempt quite similar to that of Marcel Proust, who, according to Louis Simon, sought to replace the inevitable diachrony of "life time" with the synchrony of "lived moments".[26]

The subject of trauma, with its symptoms and possible remedies, has attracted much scholarly attention in various domains of humanistic inquiry, such as psychoanalysis, Holocaust studies, genocide studies and memory studies. Among the possible reactions to loss, which according to Dominick LaCapra is a component of structural trauma, two interrelated strategies might be distinguished: melancholy and mourning.[27] Anna Zeidler-Janiszewska, referring to the Freudian concept of melancholy and mourning, argues that while melancholy is unproductive and self-orientated, mourning might be considered as a phased process of creatively working one's way through trauma.[28] The post-war cultural experience that I proposed to describe in terms of silence opens up a space for both melancholy and mourning. However, with regard to Białoszewski's "Old Life", a clear distinction between these two strategies cannot be established. On one hand, Białoszewski attempts to transcend and work through the memory of the wartime destruction, heard in the post-war silence, by replacing it with antebellum urban noises. On the other, he also acts as a melancholic, while searching for lost sounds to recreate the ambience of the ruined city in his text.

Białoszewski reacts to the post-war silence as an earwitness of the war period, as can be seen in his *Memoir of the Warsaw Uprising*, where auditory experience of the war is translated into text. One might suppose that for earwitnesses the transition from war to the post-war period involved distancing themselves from the traumatic event by expelling it from the subjective and collective awareness.

---

26 Simon, "Narrative and Simultaneity", 1.

27 Bojarska, "Trauma", 503.

28 Zeidler-Janiszewska, *Między melancholią a żałobą*, 7.

Silence seems to be a suitable category for grasping both the individual and the collective effect of such expulsion. Viewed from such a perspective, post-war silence exemplifies the specific type of veil in which trauma is hidden. Taking this into consideration, one may note that the aural specificity of this historical moment opens up a space for literary practice designed to temper memory of the evil of the war through mourning. In the work of Białoszewski, the experience of post-war silence is worked through by means of literary gestures of the recalling, repeating and recreating of lost sounds. Such a process of mourning, which in the auditory plan is based on providing sound for the post-war silence, is thus strictly connected with the literary operations performed on time.

The melancholic qualification of Białoszewski's writing might be asserted in reference to the concept of Nicolas Abraham and Maria Torok. Drawing on that concept, Andrzej Marzec distinguishes two different "compensative" strategies that might be adopted by a subject dealing with severe loss: "introjection" and "incorporation".[29] In his attempt to recreate the pre-war soundscape, Białoszewski adopts the strategy of incorporation. Marzec writes: "Incorporation is first of all a fantasy of the subject imagining that he or she is saving the dead person, who can still live in the safe place prepared for him. [...]. In that way, the melancholic subject becomes a cryptophor – the carrier of the crypt, or void, that is being looked after, nurtured and caressed within".[30] From such a point of view, the post-war silence might be imagined as a quasi-material, time-based quality that on one hand allows one to forget but on the other is constantly referring to what cannot be heard. Thus to be silent about silence would be to create a crypt, which prevents the subject from forgetting about the experienced loss.

Literature as the sound memory of a post-war city[31]

The concept of sound memory is closely connected with the concept of cultural memory. For Mieke Bal, "cultural memory links the past to the present and future".[32] Carolyn Birdsall uses the concept of sound memory "as a tool for understanding performances of past war experiences in the present, and establishing these acts as (re)constructions of the self in relation to others and to place".[33] In the case of sound memory, the practice of linking mentioned by Bal concerns aural

29 Marzec, *Widmontologia*, 88–89.

30 Ibid., 91.

31 Birdsall, "Sound memory".

32 Bal, Introduction, vii.

33 Birdsall, "Sound memory", 111.

aspects of the remembered past sounds of everyday life, as well as the sounds of speech. The city uses the textual medium of literature as an archive for lost sounds that might be brought back to hearing. Through the recollection performed by the writer, sounds from different time modes and past urban realities are relocated to the present. Thanks to this specific capacity to reconstruct past objects of experience, literature constitutes an aural mediation between past and present soundscapes. Such an observation leads one to ponder the extent to which literature changes the auditory experience of the given "heard worlds".

What is not heard any more in the real city space, causing a collective experience of absence, might be recreated in the cultural space of literature. Metaphorically speaking, literature becomes a home for lost sounds, providing a potential opportunity for them to be heard again through the practice of reading. Such a point of view coincides with Białoszewski's vision of literature, in which he accentuates the role of the spoken word. "For a couple of years now, my poems have been written not just from my imagination, but from action. I try to make what is written a record of the spoken".[34]

In this sense, literature as sound memory makes the post-war urban soundscape a haunted reality, in which the temporal orders of the past and the present overlap. The concept of hauntology was formulated by Jacques Derrida in his book on the reception of Marxism in late twentieth century Europe, published in French in 1993.[35] In this book, Derrida seeks to portray the paradoxical character of the relationship between past and present that results from its mediation by spectres. Spectres are a kind of phenomenon that cannot be attributed to the status of either past or present. They operate somewhere between the two, destabilising the framework of the concepts and categories of Western thought.[36] The Derridian concept of hauntology, which opens up the present to the spectres of the past, might be informative in studies of the auditory experience of historical soundscapes. From such a perspective, historical aural realities are not located somewhere in the past, but through the mediation of various media they are also present in contemporary auditory experience. Taking such a point of view into consideration, one may note that studies on historical soundscapes need to combine diachronic and synchronic perspectives.

If one considers literature as part of an imagined city, as I suggested above, then literary practice brings another layer of time to the physical urban fabric. Seen

34 Białoszewski, "Mówienie o pisaniu", 12.

35 Derrida, *The Spectres of Marx*.

36 Marzec, *Widmontologia*, 126–133.

from such a perspective, the post-war soundscape constitutes a multi-dimensional and contradictory object of study. By incorporating the effects of the cultural working-through of traumatic experiences, it becomes a construction that transcends linear historical periodisation. The particular historical moment of the post-war period is haunted by cultural practices of sound memory. Consequently, it consists of different times and different urban realities. In light of this, I consider that analysis of the post-war urban soundscape and of the meanings which earwitnesses attached to the auditory category of silence must be conducted together with analysis of the meanings attached to pre-war sounds.

Conclusion

In this paper, I was seeking to present the multi-dimensionality of the phenomenon of post-war silence through the interpretative excavation of meanings attached to it by Miron Białoszewski in his set of short stories "Stare życie" [Old life]. I proposed to understand silence as the earcon of absence, which makes it both an acoustic and a socio-cultural phenomenon.

The relationship between auditory experience and literature was problematised in terms of the response to auditory stimuli, the quality of which cannot be objectively described, but is rather grounded in the subjective awareness of loss. To understand Białoszewski's reluctance to openly thematise the post-war silence, I connected it with the notion of trauma. Considering literature as sound memory, I attempted to show how it links past and present auditory experiences and how, in such a process, the "heard worlds" of different times overlap, complicating the notion of the historical urban soundscape.

Bibliography

Bal, Mieke, Introduction to Mieke Bal, Jonathan Crewe and Leo Spitzer (eds), *Acts of Memory. Cultural Recall in the Present* (Hanover: University Press of New England 1999).

Benjamin, Walter, *Berlin Childhood around 1900*, tr. Howard Eiland (Cambridge: Belknap Press, 2006).

Białoszewski, Miron, "Mówienie o pisaniu" [Speaking of writing], introduction to *Poezje wybrane* [Selected poetry] (Warsaw: Ludowa Spółdzielnia Wydawnicza, 1976), 5–13.

– "Stare życie" [Old life], in *Rozkurz* [Wastage] (Warsaw: Państwowy Instytut Wydawniczy 1980), 6–26.

Birdsall, Carolyn, *Nazi Soundscape Sound, Technology and Urban Space in Germany, 1933–1945* (Amsterdam: Amsterdam University Press, 2012).

– "Sound memory: A critical concept for researching memories of conflict and war", in Danielle Drozdzewski, Sarah de Nardi and Emma Waterton (eds), *Memory, Place and Identity. Commemoration and remembrance of war and conflict* (London: Routledge, 2016), 111–129.

Blesser, Barry, and Salter, Linda-Ruth, "Ancient acoustic spaces", in Jonathan Sterne (ed.), *The Sound Studies Reader* (London: Routledge, 2012), 186–196.

Bojarska, Katarzyna, "Trauma", in Magdalena Saryusz-Wolska and Robert Traba (eds), *Modi Memorandi. Leksykon kultury pamięci* [*Modi memorandi.* A lexicon of cultural memory] (Warsaw: Scholar, 2014), 501–506.

Cage, John, *Silence. Lectures and Writings* (Hanover: Wesleyan University Press 1961).

Derrida, Jacques, *Spectres of Marx. The State of the Debt, the Work of Mourning and the New International* (New York: Routledge, 2006).

Donald, James, *Imagining the Modern City* (Minneapolis: University of Minnesota Press, 1999).

Eco, Umberto, *The Open Work*, tr. Anna Cancogani (Cambridge, MA: Harvard University Press, 1989).

Jasionowicz, Stanisław, *Pustka we współczesnym doświadczeniu poetyckim* [The void in contemporary poetic experience] (Cracow: Wydawnictwo Naukowe Uniwersytetu Pedagogicznego, 2009).

Losiak, Robert, "Cisza w doświadczeniu audiosfery współczesnej" [Silence in experience of the contemporary soundscape], in Justyna Harbanowicz and Agnieszka Janiak (eds), *Przestrzeń ciszy* [The space of silence] (Wrocław: Wydawnictwo Naukowe Dolnośląskiej Szkoły Wyższej, 2011), 57–64.

Marzec, Andrzej, *Widmontologia. Teoria filozoficzna i praktyka artystyczna ponowoczesności* [Hauntology: the philosophical theory and artistic practice of the postmodern] (Warsaw: Bęc Zmiana, 2015).

Rybicka, Elżbieta, *Modernizowanie miasta. Zarys problematyki urbanistycznej w nowoczesnej literaturze polskiej* [Modernising the city: an outline of urban issues in modern Polish literature] (Cracow: Universitas, 2003).

Schafer, Raymond Murray, Introduction to *The Soundscape: Our Sonic Environment and the Tuning of the World* (Rochester: Destiny Books, 1993).

Schweighauser, Philipp, "Literary acoustics", in Gabriele Rippl (ed.), *Handbook of Intermediality. Literature – Image – Sound – Music* (Berlin: Walter de Gruyter, 2015), 475–493.

Simon, Louis, "Narrative and simultaneity: Benjamin's image of Proust", *Studies in 20th Century Literature,* 21/2 (1997), 361–380.

Smith, Mark M., "Listening to the heard worlds of Antebellum America", in Michael Bull and Les Back (eds), *The Auditory Culture Reader* (Oxford: Berg, 2003), 137–163.

Slouka, Mark, "Listening for silence: Notes on the aural life", in Christoph Cox and Daniel Warner (eds), *Audio Culture. Readings in Modern Music* (New York: Continuum, 2006), 40–46.

Szalewska, Katarzyna, "Pejzaż dźwiękowy i praktyki audytywne czasu wojny" [The soundscape and auditive practices of wartime], *Białostoskie Studia Literaturoznawcze,* 6 (2015), 27–40.

Szymański, Witold, "Cisza w architekturze" [Silence in architecture], in Justyna Harbanowicz and Agnieszka Janiak (eds), *Przestrzeń ciszy* [The space of silence] (Wrocław: Wydawnictwo Naukowe Dolnośląskiej Szkoły Wyższej, 2011), 65–73.

Zeidler-Janiszewska, Anna, *Między melancholią a żałobą. Estetyka wobec przemian w kulturze współczesnej* [Between melancholy and mourning: aesthetics with regard to changes in modern culture] (Warsaw: Instytut Kultury, 1996).

Kamila Staśko-Mazur

Institute of Musicology, University of Wrocław

The voice of Polish Radio in the soundscape of Warsaw in 1945

Abstract: Due to political pressures and material conditions, the post-war effort to increase the radio coverage of Polish cities was to be based mainly on a network of wired radio stations. The aim of this chapter is to describe the practices of sound and listening in 1945 as captured within the context of the activities of Polish Radio in Warsaw, a city rising from the ruins. The presence of the radio's voice, as a key component of the soundscape of post-war Warsaw, is viewed from the perspective of its role in the public/private space, its material determinants, programming and reception, as well as the changes and direction to the expansion of radio coverage. Based on archive sources, including text documents of Polish Radio, radio broadcasts and written sources, an attempt was made to reconstruct both the process of making radio accessible across Warsaw in 1945 and also the reception and presence of the radio's voice in the city's soundscape. Analysis of these sources makes it possible to follow the transformation of the modes of emission of the radio's voice.

Keywords: wired radio station, Warsaw Radio, city soundscape, sound practices, listening practices, megaphone, in-home loudspeaker, radio programming

With the conference of 1939 in Łódź, the Nazi authorities limited access to education and culture in occupied Poland. Poles were forbidden from using radio sets and had to turn them in to the authorities; those who violated this new policy were sentenced to death or sent to concentration camps.[1] These regulations were still in force immediately after the war: in the first half of 1945, the Soviet army also prohibited the use of radio sets.[2] During the post-war period, there was no equipment to be had, and listening to the radio over street loudspeakers became the only way people could access radio news.

Despite the growing number of publications about Polish Radio, it seems that the impact the radio had on post-war authors and audiences, as well as the role that the everyday sonic environment played during that period, which saw the

1 See Markowska and Naliwajek–Mazurek, *Okupacyjne losy*, 13–14.

2 See Siezieniewski, "Informacja i publicystyka", 230. Dariusz Michalski provides a striking example of how the violation of this policy led to an execution in mid 1945. See Michalski, "Radio poza radiem", in Ossibach-Budzyński (ed.), *Polskie Radio*, 453–454.

restitution of the Polish radio broadcasting system, remain under-examined. This research paper is aimed at filling that gap by focussing on the practices of sound and listening in 1945 within the context of the activities of Polish Radio in Warsaw, a city rising from the ashes.

Given the post-war situation of Poland, it is necessary to pose a number of questions about the social practices related to radio's voice in the city's soundscape. What was the role of radio in public/private space? What were its material determinants? How did radio programming evolve? What was its programmatic and technological reception? In what direction did radio coverage expand and what was its social and cultural impact? Drawing on archive sources from the Records Department at Polish Television's Centre for Documentation and Programme Collections (Ośrodek Dokumentacji i Zbiorów Programowych Telewizji Polskiej SA, Dział Dokumentacji Aktowej), as well as archive radio broadcasts from the Polish Radio Archives, including memoirs and radio reportages,[3] an attempt was made to examine the development of radio broadcasting in Warsaw in 1945.[4] An in-depth study of the work of Polish Radio and its presence in the public space is complemented by a survey of the press of that period, with the emphasis on the magazine *Radio i świat* [Radio and the world], first published in July 1945 – a leading periodical devoted to radio.

One figure binding together the different threads of this paper – important events related to the role of radio in the process of the "assimilation" of the new post-war reality by the residents of Warsaw – is the pianist and Polish Radio journalist Władysław Szpilman.[5] Radio journalists recall Szpilman (and many other legendary figures of Polish Radio) – often as a living human *signum* of a return to the old order.

---

3 The story of how Varsovians rebuilt their city is accompanied by the post-war "urban" soundscape. At the Polish Radio Archives, you can listen to radio broadcasts telling the history of the city and featuring the sounds of the tram network, bridges and tenement houses being reconstructed, songs sung by Varsovians rebuilding their city, and rhyming calls made by tram drivers, etc.

4 This analysis focuses on the area within the city limits and does not take into consideration radio broadcasting centres located in nearby towns that were part of the Regional Directorate of Polish Radio Warsaw.

5 Władysław Szpilman (1911–2000) was a Polish pianist, composer, radio journalist and member of the Warsaw Piano Quintet. The author of this paper is currently writing her PhD thesis on the composer's oeuvre at the Institute of Musicology of the University of Wrocław.

1. "Something that was never there": Cable radio[6]

During the occupation, Poland lost most of its radio equipment—what remained was destroyed or taken out of the country. That partly accounts for the novel content and style of the first Polish Radio broadcast after the long break caused by the war, made from Lublin in the autumn of 1944.[7] Speaking about the "second birth" of post-war radio broadcasting in Poland, the Polish Radio historian Maciej J. Kwiatkowski says:

> Something completely new begins in Poland: cable radio. "Radio on the wire", "squawk boxes", "buckets" – or any number of other more or less dismissive names – played an essential role in our country. [...] Pre-war Poland did not want to adopt the [Soviet] model.[8]

In the devastated country, facing a wholly new political reality, cable radio, including public address systems, became a new form of communication. Initially, it seemed to be the best medium for both news and propaganda, but it was met with resistance from local communities. There was supposedly a joke among radio journalists: "I saw a man standing under a loudspeaker in Lithuanian Square".[9] People working at Polish Radio, which was then in the process of rebuilding itself, projected a fearful vision – half-joking, half-serious – of a loudspeaker squawking across an empty square, thereby testifying to the tough conditions and social practices of that period. Not only the post-war reality, but also the squawk boxes epitomise an oppressive occupation, including oppressive sound.[10] The new type of radio broadcasting and infrastructure, including the public address systems, ensured access to information during the post-war period and became an ambivalent symbol of the new life.

Back then, Warsaw (once called the "Paris of the North") found itself in a dramatic situation. Witnesses often use the metaphor of a terrifying emptiness, of overwhelming devastation.[11] In addition to accounts of the first signs of new life in the war-torn landscape, there are also recollections of patriotic actions. Despite the general ban on such manifestations, Poles continued to organise symbolic

6 Kwiatkowski, *To już historia*, 184.

7 Since August 1944, Polish Radio had operated in accordance with the agenda of the Polish Workers' Party. See Habielski, *Polityczna historia*, 186.

8 Kwiatkowski, *To już historia*, 184.

9 See *Radio i świat* (hereafter *RiŚ*), 12 (1945), 1.

10 For more information on the subversive ways in which sonic oppression was combated, including the Wawer scouting organisation's minor sabotage operations with the use of megaphones, see Kwiatkowski, *To już historia*, 158.

11 See *Spotkania z Warszawą*, radio broadcast.

parades.[12] A powerful impact was also made by "megaphone-type radio broadcasts". One of the locals joyously stated that the national anthem could be heard from the loudspeakers across the River Vistula: "It turned out that there were loudspeakers across the Vistula, and you could hear Polish being spoken from them!"[13] According to Maciej J. Kwiatkowski, although the role of a loudspeaker type of radio, as well as the traditional ether type, could not be considered trivial, initially it seemed rather ineffective:

> Back then, only a small group of people were listening to these radio broadcasts in Poland. Nevertheless, broadcasting into the ether of Warsaw played a crucial political role. This means that the city, which was devastated and destined to disappear from the map, was brought back to life and was rebuilding itself.[14]

The Warsaw radio station was the third radio station (after Cracow and Katowice) in post-war Poland; it went on air on 16 March.[15] However, it was rebuilt locally as part of an urban radio broadcasting system before the capital was moved from Lublin to Warsaw.[16]

1.1 "Back then we had nothing to work with"[17]: Material conditions

The decision to rebuild the radio broadcasting system in Warsaw was probably made after the occupying power withdrew from the east side of the city. The war damage was so extensive that restoring the radio station in its former form and location was not possible. The radio transmitter in Raszyn, the pre-war Polish Radio station on Zielna Street and the headquarters of the local radio station in Mokotowski Fort were all destroyed.[18] The decision to expand radio coverage across post-war Warsaw was made at the branch office on Śnieżna Street.

At the beginning of February, the staff of Polish Radio moved from Śnieżna Street to the Warsaw branch of the Polish Radio Company, situated in a tenement house at 63 Targowa Street.[19] That building had housed Polish Radio before the war,

---

12 See Stanisław Okęcki's recollections in *Spotkania z Warszawą*, ibid.

13 See Czajkowski (ed.), *Ślady pamięci*, radio broadcast.

14 See Semkowicz, *Rozgłośnia pracuje*, radio broadcast.

15 See Kwiatkowski, "Kalendarium"; Sawicki, *Kronika*.

16 See "Głos wolnej Warszawy", 1.

17 Kwiatkowski and Solecka (eds.), *Na warszawskiej fali*, radio broadcast.

18 Kwiatkowski, *To już historia*, 191.

19 The Ministry of Information and Propaganda was also located there. See Miszczak, *Historia, 236.* The names of the radio journalists and engineers who formed that pioneering team are commemorated in Stanisław Miszczak's monograph. See Miszczak, *Historia*, 236–251; Kwiatkowski and Solecka, *Na warszawskiej fali*, radio broadcast.

doubling as a warehouse. After the war, a tenement house was rather unsuitable for radio broadcasting. Yet it served not only as an editorial office, but also as a studio, office, sound effects workshop, sound-check space, warehouse and cafeteria.[20] The adaptation involved soundproofing some rooms with blankets and carpets, boarding up windows, and securing the necessary equipment and musical instruments. According to the Polish Radio staff, what made their work so unique was the minimal equipment and challenging working conditions. As Zbigniew Lipiński recalls:

> the atmosphere was very different in those days too; the daily news from the war front reported that the army was moving east. We all knew that that victory was coming, but at the same time we still had to cope with the privations of war. There was not enough food. We were starving and at the same time the great fanfares of victory were playing on the radio. Apart from that, we had nothing to work with. I remember writing my radio scripts on a windowsill – and that's the way it was back then.[21]

The Polish Radio headquarters functioned in an unusual way, because the city was basically an urban wasteland. The headquarters also served as a bedroom and a home to its employees, which all made professional work more difficult. There is even a famous anecdote about Władysław Szpilman falling asleep under the piano and snoring so loudly that an announcer had to silence him to continue reading the late news. Another anecdote concerns the director Roman Jasiński, who had the rare privilege of sleeping not on the carpet, but in a bathtub (which was also a director's office, buffet, etc.). The Spartan conditions went hand in hand with minimal equipment, and this all required a great deal of imagination from radio producers. Production and technical solutions applied at the radio station often involved the use of simpler alternatives to professional devices and recordings. Jerzy Kotkiewicz reminisces about an improvised signalling device:

> this crude device provided a signal to turn on a mic for the announcer. There was a box, and inside that box, there was a light bulb. [...] Once the announcer pulled the string, the light bulb automatically came on, and then I turned on the mic. Apart from that, I had a gramophone beside me and put the record on whenever the announcer finished talking. So I just had to figure out when he was going to stop.[22]

---

20 For more information on the Polish Radio headquarters on Targowa Street, see interviews with Zbigniew Rymarz, Jerzy Kotkiewicz and Marianna Sankiewicz in the series *Radio w pamięci zapisane* [Radio inscribed in our memories], http://radiopolska.pl/90lat/powojenna-odbudowa-polskiej-radiofonii, accessed 12 September 2017.

21 Kwiatkowski and Solecka, *Na warszawskiej fali,* radio broadcast.

22 Semkowicz, *Rozgłośnia pracuje*, radio broadcast.

The post-war years brought the return of the "silence filler" – a pre-war invention that was interesting, but sonically of minor importance. Back then, the sound of a clock chiming was broadcast to fill "dead" airtime. Speaking about producing sound effects in 1945, Jerzy Kotkiewicz refers to a similar invention:

> I remember that at some point we needed a kind of filler. Sometimes guests didn't have time to change places, and we had to do something to fill in anything up to a minute. So we came up with an old alarm clock. We put it into an old box (which was filled with cotton wool) – the alarm clock produced a ticking sound and worked as a perfect signal and filler of those little breaks.[23]

One of the two flats occupied by Polish Radio in the tenement house on Targowa Street served not only as a radio studio and a back office, but also as a broadcasting centre and amplifier room. Accounts provided by technicians give us an idea of the number of "subscribers" to Polish Radio:

> 250 watts – that was considerable power at the time. With every loudspeaker having a quarter of a watt, the radio station had 1000 subscribers. We were also working on deploying a broadcasting station located in a railway wagon standing on a side track behind Warsaw Zoo. [...] So basically we just needed to plug in a cable. We had a radio station that was broadcasting at close range.[24]

Nevertheless, it was public address system infrastructure that allowed radio sounds to be heard in the city's soundscape. On 11 February 1945, the first radio announcement was transmitted over the public address system after a five-year break: "Hello, hello, this is Polish Radio Warsaw". Those words were spoken by the first post-war announcer, Janina Lewandowska. The composer Zbigniew Turski had the honour of announcing the first radio broadcast. The schedule of the station's inauguration (on 11 February) included the national anthem, speeches made by the Minister of Information and Propaganda, Stefan Matuszewski, and the head of Polish Radio, Wilhelm Billig, a "poetic montage on Warsaw" presented by Roman Sadowski, and works by Chopin played by Jan Ekier and Eustachy Horodyski.[25] That is what the sources say. Yet the question remains as to the Warsaw radio station's programming and reception over the following months. Speaking about the programming of the first radio broadcasts, Kwiatkowski reminisces about the unusual conditions of reception:

---

23 Ibid.

24 See Jerzy Kotkiewicz's recollections in Zadrowski, *25 lat Polskiego Radia*, radio broadcast.

25 On Ekier, see Kwiatkowski and Solecka, *Na warszawskiej fali*, radio broadcast; on Horodyski, see Kwiatkowski, *To już historia*, 197.

> The news from the war front was obviously the most popular, but other radio broadcasts also captured the public's attention. People were standing outside braving the freezing cold, listening to the radio.[26]

1.2 "Radio on the wire": Radio infrastructure

"Wiring the city" was part of the Council of Ministers' plan (June 1945) to expand cable radio coverage across the country.[27] The reason for this decision was the difficult financial situation in Poland and the impossibility of quickly re-establishing a radio engineering industry that would produce a sufficient number of traditional radio sets.[28]

The expansion plan for the radio network in the capital city included two projects: a cable radio station in the Praga district – considered as "a separate entity for the purpose of radio coverage expansion" – and a provisional cable transmission network in Warsaw, which was part of a larger urban reconstruction plan.[29] Radio journalists reminisce about the public address system (employing a Philips radio receiver and a Russian amplifier) that was used to broadcast from Polish Radio Lublin: a set of street loudspeakers placed along Śnieżna Street (near the depot). It operated from the day when the east side of Warsaw was liberated[30] until 7 February,[31] when an order was given to disassemble the power supply. According to Jerzy Kotkiewicz, "since the radio broadcasting centre was shut down and the [Warsaw] radio transmitter was not yet installed, there was silence in the air".[32] At the same time, work was already underway on setting up a new "Praga" cable radio station, located on Targowa Street, which had all the necessary equipment installed by a team of technicians (managed by Wacław Banasiewicz).[33]

The expansion plan for the Praga radio station included three main lines: a Grochów line running south, a Targówek line running south-east and a Bródno line running north (north-west).[34]

---

26 Kwiatkowski, *To już historia*, 198.
27 See *RiŚ*, 17 (1945), 2.
28 The Philips and Tungsram factories, which produced vacuum tubes, were destroyed during the war.
29 See Wilkowicz, "Techniczny i finansowy plan", 10.
30 See Miszczak, *Historia radiofonii*, 236.
31 See Sagan, "Jerzy Kotkiewicz".
32 Ibid.
33 Kwiatkowski, *To już historia*, 197.
34 See "Projekt budowy", 1–7.

Figure 1: Site plan for Praga district cable radio station (30 June 1945, Władysław Rabęcki, Cable Radio Building Department). From Records Department at Polish Television's Centre for Documentation and Programme Collections, Warsaw[35]

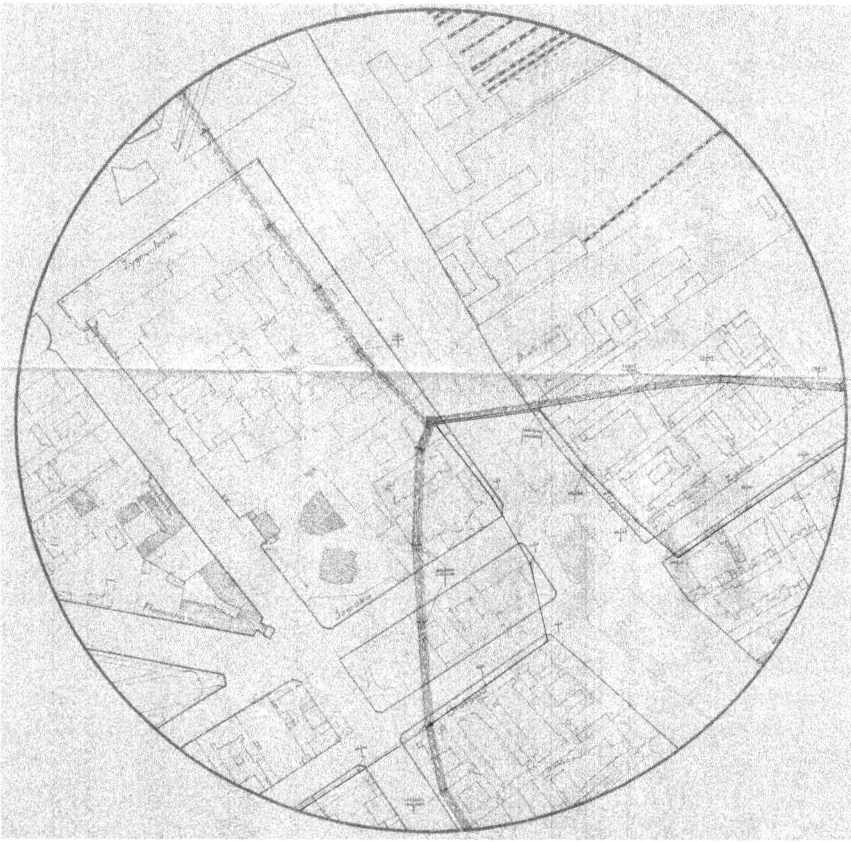

The plan of the Praga cable radio station (25 April 1945), prepared by J. Karaś, shows the infrastructure of subscriber extensions, subscriber feeders and the public address system, with triangular symbols for the loudspeakers.

35 Ibid.

Figure 2: Part of the plan for Praga district cable radio station (25 April 1945, Jerzy Karaś). From Records Department at Polish Television's Centre for Documentation and Programme Collections, Warsaw[36]

36 Ibid.

The cable radio station in the Praga district – an area then inhabited by around 300 thousand people – was to undergo a three-stage expansion over a period of three years.[37] According to the construction design for the first stage of the expansion, the distribution network was to be powered by two TU-500 amplifiers with 1000 watts of total power. The description of the state of the cable radio station over subsequent months is based on archive documents (see Table 1 below).[38]

Table 1: Aggregated data on Praga cable radio station. Prepared by the author from reports of Polish Television's Records Department. The table uses the project's original nomenclature

Praga cable radio station (east bank of the Vistula)						
Date	**Expansion plan:** Stage 1	**Expansion plan:** Further stages	Status on 25 May 1945	Status on 26 May 1945	Status on 25 October 1945	Status on 1 February 1946
Main line length	22.6 km	Remote districts and suburbs	59 km of network cable*	No data	7.2 km (23% of planned construction)	17.69 km of permanent network cable*
Subscriber line length	42 km	Concentration				
Number of street loudspeakers	35	100	32	32	20	25
Number of subscriber loudspeakers	5,000	30,000	50	123 "loudspeaker points" (50 under construction)	590	"1100" subscribers

* No distinction between main and subscriber lines

Less than two months after the opening of the Praga cable radio station, on 5 April (when the Warsaw radio station was already available over the air), a second cable

37 Ibid.

38 See "Projekt budowy", 1–7; "Stan czynnych radiowęzłów", 25 May 1945, 2 (fol. 15); Wilkowicz, "Sprawozdanie stanu robót", 28 May 1945 (describes the status on 26 May 1945), fol. 11; Wilkowicz, "Sprawozdanie według", fol. 55; "Wykaz", fol. 29 (describes the status on 1 February 1946).

radio station – the first on the east bank of the Vistula – was launched.[39] It was named "Polonia", after the hotel in Ujazdowskie Avenue where it was located. The Polish Radio Archives contain plans of the cable network, which give us an idea of the location of the two cable radio stations in Warsaw and the routes followed by their transmission cables. The plans also show a passage to the "Roma" building.[40] The main auditorium of that pre-war cinema-theatre, saved from the wartime inferno, is also mentioned in other documents that testify to its connection to the cable radio stations, and to the concert and speech transmissions that were broadcast from it.[41]

Figure 3: Plan of the cable transmission network in Warsaw (26 June 1945, Jerzy Karaś). From Records Department at Polish Television's Centre for Documentation and Programme Collections, Warsaw[42]

39 Kwiatkowski and Miszczak state that the "Polonia" had only 100 watts of power, while Polish Radio's text records mention a figure of 500 watts. See Kwiatkowski, *To już historia*, 201; Miszczak, *Historia radiofonii*, 237. Jerzy Kotkiewicz also confirms that the radio station existed.

40 Roma had a capacity of 2000. It also housed the National State Council.

41 See "Statystyka", 4.

42 "Plan kablowej sieci", 5.

Data relating to the network size, available in reports from various institutions, is not always clear or congruent.[43] The documents often use different terminology for similar information, for example: number of street and in-home loudspeakers or loudspeaker points, subscriber loudspeakers or collective connections for entire apartment buildings; the length of the network is presented in total kilometres, in kilometres per line, or per temporary and permanent lines together. An aggregation of these statistics regarding the "Polonia" cable radio station is presented, based on reports, in Table 2.

Table 2: Aggregated data on the "Polonia" cable radio station. Prepared by the author on the basis of archive sources from Polish Television's Records Department[44]

"Polonia" cable radio station ("Warsaw – City")			
Date	Status on 25 May 1945	Status on 26 May 1945	Status report from 25 October 1945
Network length	12 km	No data	17.5 km (aggregate temporary and permanent lines)
Number of street loudspeakers	16	18	22
Number of subscriber loudspeakers	2	52 "loudspeaker points" (120 under construction)	192

As shown in the reports of the cable-laying department, the installations were still largely temporary in the first half of the year and may be considered to have been *in statu nascendi*. There were complaints about the lack of network current transformers, current limiters, loudspeakers and minor installation equipment, as well as tools, professional staff and transport.[45]

In the second half of the year, depending on available materials, equipment and transportation, the plan was to build 500 more loudspeaker points, replace temporary lines with permanent ones, lay down subscriber extensions from the permanent

43 e.g. the Radio Broadcasting Office, in a report from 25 October 1945, mentions 692 in-home loudspeakers across Warsaw (minus 21 subscribers who bought lamp receivers or had poor reception), while the Cable Radio Building Department mentions 782.

44 "Stan czynnych radiowęzłów", 25 May 1945, 2 (fol.15); Wilkowicz, "Sprawozdanie stanu", 28 May 1945, fol. 11; Wilkowicz, "Sprawozdanie według", fol. 55.

45 "Sprawozdanie stanu", fol. 11.

lines and gradually switch subscribers over to them, and attach all permanent cabling to utility poles.[46] In November, there were plans to start building a trunk line towards Grochów, complete the trunk line to Saska Kępa and lay down 14 kilometres of subscriber extensions from existing trunk lines;[47] in December, the subscriber network along the Bródno and Targówek trunk line was to be extended.[48] We also know that in November an effort to promote in-home loudspeakers was made,[49] and that the alterations to the Praga cable radio station continued in December.[50]

Today, it is difficult to assess the accuracy of the reported network lengths and the degree to which the reported subscriber numbers corresponded to actual radio listeners. They may have been bigger than reported, given the complaints about so-called *pajęczarze* ("spider people"[51]) making illegal connections to the network. On the other hand, the archive documents contain notes about subscriber loudspeaker malfunctions and work orders for their repairs. For example, an inspection of 26 October shows that out of fourteen inspected loudspeakers only four worked correctly and were "fairly functional".[52]

1.3 "We have to respect tradition and the conservatism of Polish radio listeners": The policy behind the propagation of radio broadcasting[53]

The campaign to introduce radio broadcasting across the country, despite the technical difficulties, involved propagating communal radio listening through the gradual development of central cable radio stations and "collective radio devices". In a letter to the head of Polish Radio, Wilhelm Billig, of 7 August 1945, Leon Retmański mentions a number of ways to support that campaign.[54] One of them

46 See reports of the Cable Radio Building Department and work plans of the line installation and network divisions of that department, Polish Television Records Department, file no. A 85/19.

47 See Witort, "Plan prac Wydziału Budowy Radiowęzłów", 9 November 1945, fol. 54.

48 See Tomaszewski, "Plan prac działu", fol. 90.

49 Ibid.

50 Szelegejd, "Plan prac na m-c grudzień 1945r.", fol. 89.

51 For example, the report from September 1946 states that Warsaw had only 22,500 subscribers and about 25,000 "spider people". See "Raport", 1–2.

52 See Banasiewicz, "Do Naczelnego Dyrektora PR", 58.

53 Retmański, "Dot. radiofonizacji kraju", 1.

54 According to research conducted by Monika Gabryś, head of Polish Television's Records Department, Retmański was employed by Polish Radio as a production agent for a single day: 27 April 1945.

is to print posters showing radio listeners (with their families) enjoying their loudspeakers, in contrast with their *démodé* neighbours using headphones, their families tiptoeing around them. Retmański was aware of society's general dislike of cable radio, and mentioned that, due to Polish listeners' conservatism and traditions, they would have to be "nurtured for future radio broadcasting".[55] He therefore suggests "conducting a planned propaganda campaign using the press to stimulate dormant consumption in our society" and presents concrete arguments for the superiority of cable radio broadcasting over the old crystal radio sets.[56]

Eng. Edward Krok held a completely different position regarding the propagation of radio broadcasting in Poland: he criticised the wiring of the country and saw it as a retrograde step that would not be accepted by audiences, who – he thought – had different listening habits.[57] Moreover, he considered the cable system to be too expensive and susceptible to failure. Krok suggested a system based on ultra-short waves, which would ensure better reception and cheaper maintenance, enable ownership of personal receivers designed for this purpose and rid the cities of the "crosses of antennas". According to Krok, the introduction of such a system was also supported by a crowning political argument: "no one will be able to receive other stations than local ones, because VHF (very high frequency) waves have an optical range. No enemy propaganda will have any effect here, because no one will hear it". Krok argued that cable radio stations could function only in rural regions, where they would be the most effective.[58]

Discussion over the radio's mode of operation also appeared in the press, for example in the magazine *Radio i świat*, in the very first issue of July 1945. The feuilletons problematising the issue of signal carriers formed a kind of a rejoinder to the recently (26 June 1945) passed bill revoking the ban on radio receiver ownership and the ensuing discussion over the need for further "wiring of the towns and cities". Press articles fall into two groups: those that greeted the law change with ironic humour (one illustration was subtitled "Have you heard – you can

55 Retmański, "Dot. radiofonizacji kraju", 1–2.

56 Ibid.

57 See Krok, "Uwagi", 1–4. As we learn from this document, Krok worked as a cable radio supervisor in the USSR and in Proskurow Stryj, as chief building supervisor for Ukrainian cable radio in Kiev, and in German aviation, where he used VHF waves in its warning and landing system.

58 Ibid. Eng. Władysław Heller put forward an intermediate option, whereby suitable receivers would be produced and different types of broadcasting equipment would be considered. See Heller "Problem", 1–4.

own a radio now, there must be something behind it"),[59] and those adopting the tone of the official state policy. In the latter case, the authors appear to maintain an objective tone by noting listeners' understandable "mental resistance" to cable radio, resulting from the frequent use of loudspeakers by occupation forces during the war.[60] At the same time, they maintain a didactic manner and continue promoting cable radio broadcasting and loudspeakers:

> they are not meant to replace or compete with wireless radio receivers, but to complement them in order to reach full radio coverage of the country, that is, to make it a real possibility for all our citizens, rather than a select few, to get in touch with the outside world, its problems, tasks and goals, as well as its culture.[61]

There was also debate over the superiority of certain types of radio sets. Crystal radios ("cat's whisker" receivers) were criticised and discredited, despite their many advantages, because "it is unreasonable to expect anyone to spend several hours every day sitting around with headphones on".[62] So-called "people's receivers" were also criticised along similar lines, due to their heavy reliance on lamps and electricity. At the same time, there was praise for the constantly improved and modernised loudspeakers connected to cable radio stations:

> Although we do not disavow radio receivers, through crystal up to and including superheterodyne, we must conclude that cable radio is the most readily available and effective way of popularising radio broadcasting. Those who deny the utility of building cable radio stations might as well reject the idea of universal radio access in general.[63]

The tone of the *Radio i świat* columnists seems to have been carried over from the guidelines presented at the National Conference of Representatives of the Regional Directorates of Polish Radio (17–19 July 1945), where the head of Polish Radio, Wilhelm Billig, presented his report on the first year of Polish Radio. In his summary, he stated that around three hundred cable radio stations and over ten thousand home radio sets were in use across the country.[64] He also underlined that thanks to the lifting of the ban on radio ownership, Polish Radio had started registering radio devices. The number of registered devices mentioned in the

59 *Przekrój*, 13 (1945), 11.

60 See *RiŚ*, 1 (1945). For more on this subject, see Birdsall, *Nazi Soundscapes*, and Sterne, *The Audible Past.*

61 See "Odbiornik czy radiowęzeł?" [Receiver or radio centre?] *RiŚ*, 1 (1945), 8.

62 Ibid.

63 Ibid.

64 See *RiŚ*, 1 (1945), 2.

report is 100 thousand, so – in the management's opinion – about one-twelfth of the pre-war figure. Billig therefore comes to the following conclusion:

> It is clear that the ability to listen to the radio via receivers and detectors is very important and will accelerate the spread of radio access. The basic path that we chose at the very start remains valid. […] [However], for radio to become universal, we must continue to use cable radio infrastructure, which is so well supplemented by wireless receivers. Loudspeakers have numerous advantages over wireless devices due to their lower price, better availability and ease of use.[65]

Later issues of *Radio i świat* also feature articles encouraging the collective use of loudspeakers. One example is a guide to connecting various numbers of new piezoelectric speakers to different types of radio devices based on AL4 and KDD lamps.[66] *Radio i świat* continues with its praise of cable radio and criticism of private listening practices as late as the end of October:

> Given the current state of technology, the use of detector-based radios is becoming anachronistic. The sight of a man wearing headphones reminds us of the first attempts at radio reception from 1923–24.[67]

By October, an anonymous columnist, in line with Retmański's proposal to promote the loudspeaker as a status symbol of modern people, predicts that a speaker connected to the cable radio network will soon become a hallmark of a cultured person, much like such material artefacts as running water, electrical lighting and the telephone.[68]

Despite the nationwide pro-radio propaganda campaign, Warsaw still had to deal with a lack of equipment and technical issues. A report criticising the way the "Polonia" cable radio station operated, published a year later, shows that it was often impossible to meet the demand for loudspeakers, and that their placement was often less than optimal:

> After many months of work on the part of some sixty people, […] the "Polonia" cable radio station has only managed to install 736 loudspeakers. To make things worse, the loudspeakers are placed along the axis of Marszałkowska and Puławska streets, which means that mostly shops are covered, while working-class neighbourhoods like Wola and Ochota receive no service at all. The "Polonia" cable radio station has received requests

65 Ibid., 1.

66 "Wykorzystajmy każdy wat energii elektro-akustycznej" [Let's make use of every watt of electro-acoustic energy], *RiŚ*, 2 (1945), 10.

67 *RiŚ*, 14 (1945), 9.

68 Ibid.

to install over 400 more loudspeakers, which it cannot fulfil, because it only has a single 500-watt amplifier, and no loudspeakers in stock.[69]

An interesting observation is that in 1945, despite the national policy of promoting in-home loudspeakers, they did not seem common at all. The popularity of different types of receivers can be gauged from Polish Radio's reports which feature the numbers of receivers registered with the Radio's Regional Directorates. The report for September 1945 places Warsaw sixth with regard to the number of radio sets (after Poznań, Katowice, Łódź, Bydgoszcz and Cracow), with 5,040 devices within the city limits and 7,994 across the entire administrative district. What is even more interesting is the staggering domination of lamp-based devices over in-home speakers (a similar trend can be observed in the other cities).[70]

Table 3: Number of radio devices registered in Warsaw (based on data from archive documents of Polish Television's Records Department)[71]

Devices registered in the city of Warsaw in September 1945	**Lamp-based**	**Detector-based**	**Subscriber loudspeakers**
Number	3,400	283	1,357

To sum up, the quoted documents dealing with the aspirations and teething problems of Polish Radio's first year of operation indicate two important trends: the pervasive propagandistic promotion of communal listening via cable-radio-based loudspeakers, and the persisting presence of technical issues coupled with constant calls for access to working devices and mass media. The conducting of inspections of subscriber loudspeakers and the existence of sources mentioning intentional damage to wires could indicate that a part of society did not agree with the way the radio operated or with the content of its transmissions.[72] A detailed analysis of that content – especially ideological aspects – is a separate issue not raised in this article, but deserving analysis from the perspectives of sociology, political science and cultural studies.[73] Statistics showing the popularity

69 See "Raport z inspekcji w Dyrekcji Okręgowej P.R.", 2.

70 See "Sprawozdanie z działalności Polskiego Radia za wrzesień i październik 1945", 4.

71 Ibid.

72 See Bołdok, [Report], 1–2. The report describes an inspection of the wires between 63 Targowa Street and the Roma building.

73 The radio scholar Maciej J. Kwiatkowski points out that such analyses are often difficult, due to a lack of sources. See Kwiatkowski, "Jerzy Myśliński", 148.

of different kinds of receiving devices seem to be at odds with official state policy. According to those figures, lamp-based radios were the most popular (despite their alleged high cost), probably owing to the fact that they allowed users to freely choose which station they listened to.

2. "Socialist in content, pre-war in form": Radio programming

In assessing the first stage in the rebuilding of Polish Radio, Maciej J. Kwiatkowski paraphrases a famous adage and calls Polish Radio between 1944 and 1949 "socialist in content, pre-war in form".[74] The development of post-war radio programming progressed along with the expansion of the team and technical advances, and in accordance with the changing political regulations.[75] In general, the programming framework, despite the extremely difficult working conditions, clearly shows a desire to reflect the pre-war diversity of broadcasts in both their musical and spoken components. Let us remember that during the inter-war period Polish Radio's programming framework developed fully, presented a complete range of broadcasts, divided into summer and winter programmes, maintained good proportions between words and music, and was differentiated between day and night. Beginning with the signal and morning gymnastics, there was news and music, "radio mailboxes", live and recorded music, transmissions from dance clubs and revue theatres, reports and other coverage, which grew even more popular after the introduction of recording equipment capable of using Neumann's soft discs and Stille's tape. Radio plays and "readings" were also featured, and, up to 1939, occupied third place percentage-wise after news and music.[76]

According to sources, the early cable radio broadcasts were often characterised by an uneven, ad hoc structure dominated by news programmes and music (either live or played from the few available records). I propose an analysis of these broadcasts as mentioned in historical accounts.

---

74 See ibid.

75 Radio broadcasting, rebuilt after the war, was strictly censored: from 19 January by the Central Office for Control of the Press, Publications and Entertainment attached to the Ministry of Security; from 19 November by the Main Office for Control of the Press, Publications and Entertainment. See Adamowski, "Od radiofonii", 33.

76 Kwiatkowski, *"Tu Polskie Radio Warszawa"*, 178–235.

2.1 "A station's signal is its calling card"[77]

One of the first important signals accompanying Polish Radio since its inception was the "Dąbrowski Mazurka",[78] which signalled the end of broadcasting at midnight from 1926 onwards. Presented on air at moments that required an appropriate setting and national symbolism, the Mazurka (from 26 February 1927, the Polish national anthem) found a permanent place in the programming, and – as already mentioned – it was the first tune heard from the loudspeakers "across the Vistula river" after the war. The radio listing guides available in newspapers show that the anthem was played twice every day: at the start and the end of the day.[79]

What ended up being the real signature tune of the Warsaw radio station were the first few bars of Chopin's Polonaise in A-major op. 40 no. 1, first played probably on 21 December 1927,[80] proposed and performed by the radio pianist and accompanist Ludwik Urstein.[81]

After Chopin's Polonaise became the emblem of the Polish radio station, it accompanied the Warsaw I radio station up to the death of Marshal Józef Piłsudski. It was then, after the period of mourning (which, in the phonic dimension, was marked by a drumroll motif) and a change in the management at Polish Radio, that, to a request by Karol Krzewski, head of the Main Programming Board, the radio's signal was changed to an excerpt from the legionnaires' song "My Pierwsza Brygada" [We are the First Brigade].[82] That song was recorded by the then newly employed pianist Władysław Szpilman.[83] The signal, although initially planned

---

77 See F. P., "Zaczynamy", 4.

78 A Polish patriotic song by Józef Wybicki. Kunert (ed.), *Mazurek*, 21–37.

79 See guides available in the press.

80 Paweł Sztompke mentions this date in his radio audition "Spór o datę narodzin Jedynki" [The dispute over the date Polish Radio 1 was born], http://www.polskieradio.pl/7/160/Artykul/588353,Spor-o-date-narodzin-Jedynki, accessed 13 September 2017.

81 See Kwiatkowski, *Tu Polskie Radio*, 126.

82 Ibid., 239, 286. "We Are the First Brigade" was written at the end of the First World War, with lyrics by Andrzej Hałciński and Tadeusz Biernacki and music based on "Kielcki marsz nr 10" [Kielce march no. 10]. In 1924. Marshal Józef Piłsudski named it the official song of the Polish Legions, which boosted its popularity and helped it achieve the status of a semi-official anthem of the reborn Polish Army. See Kunert, *Mazurek*, 74–78.

83 Here is the head of Polish Radio, Andrzej Siezieniewski: "Szpilman started working for Polish Radio on 1 April 1935, at 24 years old, and became a part of its history on 12 May 1935, when the station's signal was changed after the death of Marshal Piłsudski. Szpilman played and recorded the main theme of 'We Are the First Brigade' on tape".

"for one year" only,[84] stayed on air from 23 June 1935 (8:45 p.m.) to 6 September 1939, when both radio masts in Raszyn were blown up. It was on 7 September that the Warsaw II radio station started its broadcast with the first beats of the song "Warszawianka".[85]

We can assume that during the first month of the Warsaw radio station's operation, in 1945, the signal was played live by an on-duty pianist. Radio employees remember it being performed.[86] One account mentions it within the context of the difficulty in obtaining a professional piano for the radio and the instrument's unusual use:

> the piano and Władysław Szpilman served as the station's signal [...] and its central office. The office employed one person – Konrad Zawadzki – and was located on a suitcase at the piano's tail-end. Over the [piano's] black case, a staring contest raged between the artist, who was being disturbed by the clerk, and the clerk, who was being disturbed by the artist. Relicts of that antagonism live on even today.[87]

In the following months, the station's signal was performed mechanically, and its broadcast was in the hands of the announcers. Planned for July was "the preparation of a project for a sound signal for Warsaw Radio and the start of its creation"; the signal was to be launched in August.[88]

See "Siezieniewski: Ważne". This was also confirmed by Szpilman's wife, Halina, in a conversation from 21 March 2017.

84 F. P. "Zaczynamy" (p. 4) and *Biuletyn Radiofoniczny*, 24 (1935), quoted in Kwiatkowski, *Tu Polskie Radio*, 489.

85 Kwiatkowski, *To już historia*, 130. "Warszawianka" is a patriotic song written in Paris by Casimir Delavigne (author of many patriotic elegies) on the occasion of the November Uprising. Translated by Karol Sienkiewicz and set to music by Karol Kurpiński, it quickly became one of the most popular patriotic songs in Poland. See Kunert, *Mazurek*, 58–59.

86 According to Janina Lewandowska, "the station's signal was always played by Szpilman". See Semkowicz, *Rozgłośnia pracuje*. According to Jerzy Kotkiewicz, the post-war signal was taken from the first bars of Chopin's Polonaise in A major. See Nowicki, "Polskie Radio".

87 Kwiatkowski and Solecka, *Na warszawskiej fali*.

88 Sommer, "Sprawozdanie", fol. 21 a.

2.2 "Make something out of nothing": The cable station's musical and spoken broadcasts

During the initial period, the main components of radio programming, as supervised by Colonel Stanisław Nadzin, were information from the frontline, military and political news, and programmes about the post-war reality.[89]

The music department, headed by Roman Jasiński, attempted to present at least one live concert a day. The cycle was inaugurated on 19 February with a performance by Władysław Szpilman. Subsequent days featured performances by Jan Ekier and Jan Gorbaty, as well as the violinists Irena Dubiska, Stanisław Tawroszewicz, Zygmunt Lednicki, Tadeusz Zygadło and Igor Iwanow, among others. Szpilman also recalled organising concerts by such singers as Wiktor Brégy, Tadeusz Zakrzewski and Stefan Witas, as well as rehearsals of a dance orchestra.[90] The live music was supplemented by a set of about fifteen albums recorded in Moscow by the Polish Folk Orchestra, including just a few folk songs (obereks, mazurkas and kujawiaks).[91] Witnesses recall that the few albums were played in alternation around the clock, which required a large dose of patience from the listeners. According to the directors' memoirs, just three musicians were responsible for playing albums: Roman Jasiński, Witold Lutosławski and Władysław Szpilman;[92] the three musicians also tried to buy more albums whenever possible.[93] Many memoirs from that period, beside expressions of general enthusiasm and creative involvement, contain an obvious, although rarely acknowledged, basic observation:

---

89 Stanisław Nadzin first served as head of the Military Department, then was promoted by Wilhelm Billig to Programming Director. See Kwiatkowski and Solecka, *Na warszawskiej fali.*

90 See Semkowicz, *Rozgłośnia pracuje.* Szpilman took on a management role in February 1945. Employed after the war by Zbigniew Turski, he spent the first week as head of the Music Department; then, after Roman Jasiński (appointed Music Director by Wilhelm Billig) arrived from Lublin, he became head of the Light Music Department. See Strycka, *Wspomnienia.*

91 Recorded by the Polish Patriotic Association in the USSR. See Kwiatkowski and Solecka, *Na warszawskiej fali.* Jerzy Kotkiewicz also mentions other compositions, such as "Rota" [The oath], songs by Moniuszko, and Harris's "Piosenka o mojej Warszawie" [A song about my Warsaw], which was at one point banned, and also the practice of looking for musical literature on open-air markets. See Sagan, "Jerzy Kotkiewicz".

92 Jasiński's account gives one of the earliest dates of the start of Witold Lutosławski's work at the post-war Polish Radio. Documentation shows that he was assigned to the III-a salary group in his post of Classical Music Director. See "Zarządzenie".

93 See Jasiński's recollections in Semkowicz, *Rozgłośnia pracuje.*

> People who haven't lived through it can't imagine what it meant to create such a radio station. […] Normally, in the rest of the world, building a radio station takes years. Here we had to create something out of nothing immediately. The conditions were terrible.[94]

A few days after the station was officially opened, on 18 February, with the arrival of Wanda Tatarkiewicz-Małkowska, her first children's programme was broadcast. Daily programming was also enriched with literary segments edited by Henryk Ładosz, Aleksander Maliszewski and Roman Sadowski. Radio theatre also enjoyed a resurgence, under the eye of Bronisław Dardziński.[95] An account of the radio's full schedule, including its famous radio plays featuring Józef Małgorzewski, can be found in Pola Gojawiczyńska's serialised radio novel *Stolica* [The capital city]. Here is Gojawiczyńska on the radio station's album repertoire:

> The studio's musical assets consisted mostly of German albums, which were rather hard to come by. The cheerful tones of "The Blue Danube" often shook the entire storey, and the loudspeakers outside passed them on farther.[96]

One of the best-remembered spoken word broadcasts was the "family search mailbox". As a genre, this alluded to the pre-war radio mailbox,[97] although the programmes from 1945 had a very specific formula,[98] as described by the announcer Janina Lewandowska:

> We received massive numbers of letters from people looking for their families, so we started the mailbox. The ones I remember the best are the family search mailboxes, because they were the essence of human suffering, of tears which, in these few short lines, would overflow from the page: "Where is he, in Sachsenhausen, in Murnau, in some other camp?" Also the searching parents and grandparents. For example: "Płóciennik Tadeusz, born in 1925, wounded around Stawki at 7 in the evening, last seen near the sewer on Plac Krasińskich, wounded and carried away in an unknown direction on the day before capitulation, is sought by his mother".[99]

---

94 Semkowicz, *Rozgłośnia pracuje*.

95 Kwiatkowski, *To już historia*, 198. Anecdotes about realistic radio dramas and the special effects used in them occasionally appear in historical broadcasts of Polish Radio.

96 See Bocheński, *Znam*, 124–25.

97 Created by Marcin Stępowski; different kinds of mailboxes were popular before the war.

98 For more on the passing of information to mailboxes via a special system of hooks available at a butcher's next to the radio studio, see Kotkiewicz's recollections in Sagan, "Jerzy Kotkiewicz".

99 Zadrowski, *25 lat Polskiego Radia*.

Another announcer, Lucyna Niedziałkówna-Przybylska, also mentions the mailboxes as a key part of the daily programming, and refers to another important feature on the night schedule: news from the "Polpress" Polish Press Agency:[100]

> Back then, Polish Radio had daytime and night-time programming. Everyone remembers the daytime programming. Its focal point was the family search mailbox. Everyone was looking for everyone. Apart from news and some basic auditions and recitations, we also had a bit of music, lively music… polkas, obertas, marches. But the night-time schedule was probably more important: it was about the victorious march of the soviet army towards Berlin. We, the announcers, would read this so-called Polpress news from midnight till six in the morning.[101]

2.3 Warsaw Radio programmes in the ether

One month after the programmes began being played through the city's infrastructure, it became possible to start sending Warsaw's signal out into the ether. This happened on 16 March, when a post-war radio transmitter – often called a "Bee", due to its original use as a signal jammer – was brought to Warsaw from Lublin.[102]

The sound dimension of the station's ceremonial opening featured several symbolic elements.[103] The programme included a piano recital by Władysław Szpilman, who was the last musician to perform live on air before the war, when radio broadcasts ceased on 23 September 1939, due to the destruction of the radio transmitter.[104] The pianist also had the honour of performing a musical programme during the first on-air broadcast of the Warsaw radio station. Szpilman often recalls his performance, which lends a symbolic frame to the period of Polish Radio's activity disrupted by the war:

> Major Nadzin and director Jasiński asked me to play half an hour of Chopin, and I must admit that I was very happy that I, the one who played for the last time in September, would be the one to play at the opening […] for me, it was a very important day in my life.[105]

Another extraordinary feature was a feuilleton by Wacław Rogowicz titled "Zerwana pamięć milczenia" [The broken memory of silence], which commemorated the fallen and murdered employees and contributors of Polish Radio, honouring their memories by interrupting the broadcast with several seconds of silence.

100 For more on Polpress, see Habielski, *Polityczna*, 179.
101 See Kwiatkowski and Solecka, *Na warszawskiej fali.*
102 *RiŚ*, 5 (1945), 1.
103 The programme is given in Semkowicz, *Rozgłośnia pracuje.*
104 See Szpilman, *Pianista*, 24; also Strycka, *Wspomnienia.*
105 Semkowicz, *Rozgłośnia pracuje.*

Maciej J. Kwiatkowski notes that this was probably the first time that a Polish Radio broadcast was interrupted in this way.[106]

When the next step in Polish Radio's development – the ceremonial opening of the radio tower in Raszyn and of the new offices on Ujazdowskie Avenue – took place on 19 August, the national programme was overhauled, and the "Bee" emitter took over local broadcasts. The refreshed national programme included the religious song "Kiedy ranne wstają zorze" [When the light of dawn arises] – a symbolic allusion to the pre-war radio tradition.[107] As we know from letters sent to the radio station, the song's presence was not without significance.

The opening of the new offices was also accompanied by special programming, including a ceremonial concert and unusual "attractions". A concert by Zofia Rabcewicz included Chopin's Polonaise in A major op. 40 no.1 and Piano Concerto No. 2 in F minor, while Eugenia Umińska and Władysław Szpilman performed Karol Szymanowski's "Źródło Aretuzy" [The spring of Arethusa] and a Mazur by Aleksander Zarzycki. The second part of the concert featured Juliusz Zarębski's Quintet, followed by some popular music, accompanied by excerpts from operas, as well as pieces composed specially for the occasion.[108] For the first time since the war, the audience visiting the radio building had the rare chance to watch a radio play being performed live and to look behind the scenes of a radio production.[109]

From September, besides the Small Orchestra conducted by Stefan Rachoń, which carried on the pre-war tradition of the Radio Orchestra led from 1935 by Zdzisław Górzyński,[110] the radio boasted several new bands, including the Polish Radio String Quartet, the Radio Revellers Quartet and a jazz orchestra directed by Władysław Szpilman.[111] New broadcasts included *Opera wyobraźni* [Opera of the imagination].[112]

---

106 Ibid.

107 This composition first appeared in Polish radio broadcasting on 12 June 1933, opening the day's programmes; beginning the day with a religious song was highly unusual – normally the sole preserve of the Vatican. A recording of the song is available on the Polish Radio website, http://www.polskieradio.pl/231/4360/Artykul/1400761,Kiedy-ranne-wstaja-zorze-Glos-kosciola-w-Polskim-Radiu, accessed 13 September 2017. Poland was also one of the first countries to transmit a holy mass. See Krzysztof Górski, "Audycje religijne", in Ossibach-Budzyński (ed.), *Polskie Radio*, 287.

108 Władysław Szpilman's composition titled *Wiwat Raszyn* [Long live Raszyn].

109 *RiŚ*, 15 (1945), 8.

110 Kwiatkowski, *Tu Polskie Radio*, 350.

111 *RiŚ*, 7 (1945), 5. The orchestra's signal is available in the National Digital Archives.

112 A series of operas played from records was launched by Roman Jasiński with Bizet's *Carmen*, performed by the Milan Opera. See *RiŚ*, 5 (1945), 10.

According to archive documents, by the end of 1945 there were about sixteen hours of broadcasts per day,[113] including transmissions from Łódź, Katowice and the Roma hall in Warsaw. The broadcasts included a nationwide programme, starting in the morning (between 6.45 and 8 a.m.) with the national anthem and running until midnight, as well as local programming.[114] Percentage-wise, the nationwide programme featured mostly news; musical broadcasts (records, transmissions and retransmissions, depending on the day of the week) came second;[115] musical-spoken programmes (e.g. a weekly magazine) came third; special broadcasts (such as children's shows, or literature and poetry programmes lasting a quarter of an hour) placed fourth. These percentages are shown in a table based on statistics from Polish Radio's Research Office.[116]

Table 4:. Percentage of airtime devoted to different types of broadcasts. Based on files from Polish Television's Records Department

Type of broadcast	**Percentage of airtime (100%)**
Music	37.6%
News	23.9%
Family search mailboxes and Polish Red Cross mailbox	12.6%
Special interest	8.6%
	Including
	3.7% for Poles abroad
	3.2% for the military
	0.5% for clubrooms
Lectures, speeches, current affairs	8%
Literary programmes	4.8%
Programmes for children and youngsters	2.8%
Mixed programmes	1.7%
Programmes for rural regions	1.2%

113 "Statystyka programowa", 4.

114 A Warsaw chronicle, recorded music, feuilletons, reportages, news and announcements were played. The evening block featured foreign language programmes, as well as programmes for Poles abroad.

115 This could be a song, a violin or piano recital, a programme in the *Nasze pieśni* [Our songs] series or, in the afternoon, a concert by the Polish Radio Small Orchestra conducted by Stefan Rachoń, a light music concert or an instrumental recital.

116 See "Statystyka programowa", *Biuletyn Biura Studiów*, 3 (1945), 4–6.

According to Miszczak, in 1945, the nationwide programme consisted of 2,747 hours of broadcasts. By comparison – according to the author – the number grew to 4,702 hours in 1946.[117] This clearly shows that the amount of airtime increased systematically.

3. "Next to the atom bomb, the most fearsome invention of the twentieth century was cable radio"[118]: The practices of radio participation

Maciej J. Kwiatkowski, commenting on the first broadcasts of Warsaw Radio, considers that this kind of transmission, besides its limited reach, had the unique advantage of being able to receive instant and accurate feedback from listeners:

> never again will radio workers have a such a direct way to research their listeners' reactions, tastes and preferences. All you have to do is go out onto the balcony and look at the listeners gathering under the loudspeaker.[119]

He also mentions that the radio was an emotional subject not only for the announcers and musicians who started working there after the war, but also for the listeners:

> Tadeusz Bocheński […] had a lot of trouble getting to Targowa […] so as soon as his familiar voice sounded from the loudspeakers, the listeners on the street greeted him with applause.[120]

The legendary announcer Tadeusz Bocheński recalls the moment he returned to the radio, comparing it to a symbolic renewal of life:

> I can't express the feeling with which, after such a long break, after those dark, terrible years of occupation, I stood in front of the microphone. When I realised that I was again talking to my radio friends, scattered all around Poland, that people were listening to me in the most distant reaches of the country, or even abroad, that for them I might be a symbol of the capital city rising from the ashes, of new life awakening, I was overcome by emotion. I couldn't speak. I remember needing quite a while to get my bearings and go on speaking. That is what my first broadcast after the war was like.[121]

Apart from the understandable emotions caused by the return of a familiar voice to the Warsaw sonic environment, there were also other feelings that accompanied the megaphone and loudspeaker transmissions – feelings related mostly to their

---

117 Miszczak, *Historia radiofonii*, 331.
118 Minkiewicz, "Co słychać w Łodzi", 12.
119 Kwiatkowski and Solecka, *Na warszawskiej fali*.
120 Semkowicz, *Rozgłośnia pracuje*.
121 See *Spotkania z Warszawą*.

technical failings and the inhabitants' new practices. Several issues of *Przekrój* feature satirical commentary on the cacophony of tenement houses (caused by many inhabitants listening to the radio at the same time) or the general "pollution" of the city's soundscape:

> Inarticulate growling and unbearable static can often be heard in many apartments, institutions and waiting rooms. This is a sign that the most fearsome invention of the twentieth century is functioning: the so-called cable radio.[122]

Similarly satirical and critical opinions can be found about the transmissions played over the city's public address system. One particularly shocking recollection deals with the first post-war concert transmission from the Warsaw studio, which took place at the end of March (and can be considered representative of this period's concert broadcasts). The performance features the excellent singer Ewa Bandrowska-Turska, popular even before the war. Her show, received not live, but over a loudspeaker, elicited entirely opposite aesthetic reactions:

> Having found out about the transmission, I stood for a while under a public announcement loudspeaker but [...] to this day I'm not sure whether I heard Bandrowska or not. The sounds that reached my ears made me think of a gargling hippopotamus suffering from the flu.[123]

Nearly half a year later, a columnist noted with some optimism that by that time (in September) "*huge* progress has been made in that area". He considered that the transmissions were improving technically every day, and some of the concerts transmitted from Łódź to Warsaw II (between 10 and 11 p.m.) could "*occasionally* be enjoyed without any disturbance".[124] The author also notes that reception is diminished by the habit of playing concerts half consisting of pre-recorded material. Instead of such "double playing" of albums, he asks for the same recordings to be played with "exemplary clarity" from the home studio.[125]

---

122 Minkiewicz, "Co słychać w Łodzi". Although the commentary deals with the soundscape of Łódź, it probably holds true for many urban experiences of cable radio reception.

123 J. W., "Krytykujemy: Muzyka w eterze" [We criticise: music in the ether], *RiŚ*, 9 (1945), 5.

124 This concerns radio broadcasts that could only be heard on home radio sets. Following an intervention from radio producers, in September an instruction was issued banning the transmission of artistic programmes over street loudspeakers. However, that ban was not unconditional and was probably lifted on many occasions. Emphasis added by the author.

125 J. W., "Krytykujemy", 5.

A month later, Jerzy Waldorff alludes to the phenomenon of concert transmissions in a short review of a concert accompanying the return of Fryderyk Chopin's heart to his home country. In reviewing the Chopin pieces performed by Bolesław Woytowicz, Zofia Rabcewicz, Jerzy Żurawlew and Jan Ekier, he seems to have a positive view of the ability to listen to concerts over street loudspeakers. This time, the author is not concerned with the technical potential of loudspeaker broadcasts, but rather with how they benefit the public by playing desirable music. Waldorff sees the celebration of Chopin, as well as its transmission over street loudspeakers, as indicative of the start of some aesthetic and programmatic changes in the listening habits and practices of the local population:

> you had to see the people standing under the street loudspeakers and listening to Chopin with as much as attention as they used to pay to news about the war to understand that the Chopin celebrations were a vital part of changing Warsaw's psyche from wartime to peacetime.[126]

So how did the social practices and expectations regarding the presence of the radio's voice in the city's soundscape actually change in Warsaw's inhabitants? Did they evolve along with the rebuilding of their lives, did they follow politics, which issued guidelines regarding form and content, or did these changes occur – and if so, then how much – at their own pace? The head of Polish Radio offered some guidelines as early as August:

> Now we should move away from catering to random passers-by and concentrate on listeners who tune in from their homes and their clubrooms; on listeners who are not satisfied with just information, but who would also like to listen to good music, literature, drama and lectures, learn foreign languages – in short, on people who want radio to give them not just information, but also a chance to be educated and culturally entertained.[127]

3.1 "The programming will have to adjust to the time of peace": Changes in programming and reception

Programming changes were announced at the same time as the plan to increase the number of potential listeners. On 22 July, the head of Polish Radio, Wilhelm Billig, said the following in a speech:

> In the first year of its existence, Polish Radio paid special attention to technical questions, because it was aimed at listeners on the streets. It was a typical wartime broadcast, with all its advantages and drawbacks. News of the war was considered the most up-to-date

126 *RiŚ*, 15 (1945), 9. For the issue of staging the sound in text see: Bijsterveld, *Soundscapes of the Urban Past.*

127 *RiŚ*, 5 (1945), 2.

> and most interesting.[128] [...] Because the war has ended, the programming will have to adjust to the time of peace and work, and from broadcasting aimed at people on the street to programmes made for people listening at home.[129]

Stanisław Nadzin clarifies the direction and timeline of the reforms in his report:

> Over the past three months, we have laid only the foundation for a new, extended, changed programme adjusted to the needs of peacetime construction. The final and crucial changes in that domain are still just plans. They will be realised on the day Polish Radio's central station in Raszyn is opened.[130]

In line with the announcements, the radio station in Raszyn opened soon after, and Polish Radio started publishing listening guides and articles explaining its development plans in the press. Those articles often stressed the structured nature of the new programming, which ought to be listened to intentionally rather than accidentally (as was the case during the war): "The listeners have to be directed towards a lasting covenant with the entire programme".[131]

In order to get the best results with regards to programming and reception, listeners were encouraged to correspond with Polish Radio and send in their opinions and comments about the broadcasts. The campaign resulted in 454 letters being sent by listeners in January 1946, including 380 regarding programming, 35% of which were negative, less than 16% positive and 49% of which contained suggestions.[132] The letters, which were carefully reviewed by a special unit of the Polish Radio Research Office, give us some idea of how the programming was received by listeners – at least those who wrote in. The critical opinions including the following: "Three-quarters of the programmes consist of notices, family searches, talks about economics and many other broadcasts that cannot be considered interesting"; "there is too much preaching and politics, and not enough humour and entertainment". One example of a positive opinion was as follows: "the programming is becoming nicer, richer and more interesting".[133] With regards to music, some opinions both criticised and praised the same broadcasts. Negative opinions usually dealt with classical music: "Rural regions complain about the excess of classical music, as well as piano

128 Billig is talking about the nine-month period starting in August 1944. See *RiŚ*, 2 (1945), 1.

129 *RiŚ*, 1 (1945), 1. On the subject of staging sounds in texts, see Bijsterveld, *Soundscapes*.

130 *RiŚ*, 2 (1945), 1.

131 *RiŚ*, 11 (1945), 1.

132 A summary of listeners' letters from January and February shows only a minor prevalence of negative opinions, with 198 positive and 121 negative. See "Analiza", 12–13.

133 Ibid., 16–17.

and song recitals".[134] Suggestions included requests for more light, entertaining and dance music, while classical music lovers asked for more symphonic concerts, more Chopin and more concerts with running commentary. There are also requests for more lectures – sometimes with topic suggestions. Listeners sharply attacked the excessive propaganda and complained about being bored by political speeches and transmissions from official celebrations and rallies: "I usually turn the radio off to avoid listening to propaganda and irritating lectures".[135]

The listeners who wrote in to the radio leave little room for doubt regarding their reception of the work of field reporters. Negative comments (from as early as the beginning of 1946) about radio reportages – e.g. "sound reportages from factories and so on are huge pieces of kitsch. Perhaps if they were done differently, they would be good; right now, they're a waste of time" – were replaced by positive ones in a matter of months.[136]

Listeners' letters also reveal their education and taste. Some criticise the announcers' unclear speech, incorrect accentuation and reading as if they had no time to familiarise themselves with the text beforehand: "I would like to point out that constant errors in announcing the programme and stating the time, as well as the frequent 'goddamns' etc. are not pleasant".[137] Praise is often given to children's shows; there are even suggestions that they be broadcast before noon, as was the case before the war.[138]

A survey showing the preferences and practices of people's listening to the radio at home across the nation is another interesting source.[139] A question about the respondents' favourite programmes yielded the following result: light music 63%, radio plays 57%, light music and spoken programmes 57%. The next question, related to the radio guide ("Do you usually listen to programmes selected from the printed radio guide?"), yielded 53% of positive replies. The most interesting question with regards to the purposeful selection of programmes, worded as "Do you turn on the radio randomly?", yielded only 13% of positive replies. The last question – "Is the radio always on when you are at home?" – was answered in the affirmative by only 23% of the respondents (9% gave an indeterminate answer).[140] As the survey shows, the listeners – whatever their musical tastes – made conscious

---

134 Ibid., 18.
135 Ibid., 20.
136 Ibid., 21.
137 Ibid., 22.
138 Ibid., 23.
139 "Wyniki", 43–46.
140 Ibid., 46.

choices regarding the time and mode of listening to the radio and did not enjoy having the radio constantly playing in the background. In their comments about the quality of the broadcasts, they often proved faithful to the pre-war standards of announcing and sound engineering.

Subsequent years would bring more strategies and plans regarding the propagation of radio broadcasting in Poland's urban and rural regions. Speeches made by officials often share the same optimistic tone that prognosticates the popularisation of the new mode of listening to the radio:

> it has to be stressed that the further development of cable radio broadcasting seems certain, since the Polish public has rid itself of all prejudice – unwarranted prejudice – against loudspeakers.[141]

Despite that, an issue of the Research Office bulletin from 1946 features a less optimistic report on the direction of the national policy: "The increase, in percentage terms, of lamp-based receivers is about equal to the increase of in-home loudspeakers. This indicates that the propagation of radio broadcasting has slowed compared to the previous period, when the rate of increase was higher".[142]

At the same time, according to official guidelines, the development of the system of street loudspeakers was to be discontinued from 1946. They had served their purpose, and their usefulness was regarded as negligible.[143]

Despite that, the loudspeakers continued to function in cities, institutions and workplaces, and the personal attitudes of listeners towards the formula and quality of the broadcasts was often more complex. Apart from the cited examples of complaints about the omnipresent voice from the cable radio, its technical failings and the programmatic and propagandistic abuses, there are also some accounts of attempts to go beyond its material determinants. Some of them include surprising – given the formula of the broadcast – recollections of experiencing the discovery of musical meanings, which at the same time show the incredible perceptive and cognitive potential of listeners accustomed to the megaphone-like mode of sound reproduction. The writer Zofia Nałkowska recalls her personal experience of finding the value of music in the noisy blathering of the "Polonia" cable radio:

> On the corner of the former Marszałkowska Street and the former Jerozolimska Avenue, there are fruit- and even flower-stands. You can hear music – exceptionally beautiful for that time – from the loudspeaker: Debussy, Schubert, played by some Wanda Mossakowicz. The loudspeaker is "speaking with its mouth full", making noise and booming, but

141 See "Z frontu", 13.
142 See "Rozgłośnie", 11.
143 See "Techniczny", 3.

> the sound quality rises above it all. I am shaken, happy, ready to cry, to love and despair. I am pretending to stand in front of the flower shop window to furtively keep listening.[144]

Soon, the archives begin to turn up documents mentioning the need to respect silence and attempts to combat the familiar practices of listening by aiming loudspeakers (placed, for instance, on windowsills) towards public spaces. Here is a report from 1946: "May saw the launch of a general press campaign against playing the radio in the early mornings and late evenings".[145] The press also started carrying articles that criticised the behaviour of radio listeners.[146]

The following year, the question of the urban soundscape (in relation to the approach to promoting classical music) was addressed by composers and musicologists at a convention of the Polish Composers Union (PCU).[147] According to the "manifesto" put forward during the PCU's first convention, multiple campaigns were needed to "harness the huge possibilities offered by radio broadcasting in order to widely propagate musical culture", among other things. Aware that radio was a "factor contributing to listeners' receptive abilities", the delegates at the third convention decided to unambiguously formulate some recommendations with regard to the playing of classical music over street loudspeakers:

> We should strive to guarantee a sufficient framework for the reception of radio programmes, so that the most serious musical works are not reproduced in conditions inappropriate for the gravity of art, for example over street loudspeakers.[148]

The constant presence of the radio's voice (and its musical content) in the public and private space of the inhabitants of towns and cities therefore becomes a question of negotiating the right (both personal and public/institutional) to phonic space. It also seems to confirm Janusz Korczak's opinion that "every step changes the gravity of Earth. Radio will change the gravity of man".[149]

tr. Agata Klichowska

Bibliography

Adamowski, Janusz W., "Od radiofonii Rzeczypospolitej Drugiej i Pół do radia publicznego (zarys dziejów Polskiego Radia latach 1944–2013)" [From the radio broadcasting of the Second and a Half Republic to public radio (an outline

---

144 Entry of 23 October 1945. See Nałkowska, *Dzienniki*, 108.
145 See "Radio", 30.
146 Ibid.
147 Third General Convention of the Polish Composers Union, 20–21 October 1947.
148 See "Protokół", 3.
149 *Radiofon Polski*, 51 (1926), 4.

history of Polish Radio 1944–2013)], in Andrzej Ossibach-Budzyński (ed.), *Polskie Radio. Historia-program-technika. 90 lat Polskiego Radia* [Polish Radio: history, programmes and techniques. 90 years of Polish Radio] (Warsaw: Polskie Radio SA, ASPRA-JR, 2015).

"Analiza korespondencji od radiosłuchaczy" [Analysis of letters from the audience], *Biuletyn Biura Studiów,* 2/5 (1946), Polish Television Records Department, file no. A 1457/2.

Banasiewicz, Wacław, "Do Naczelnego Dyrektora PR" [To the Head of the Polish Radio], 31 October 1945, Polish Television Records Department, file no. A 85/19.

Bijsterveld, Karin (ed.), *Soundscapes of the Urban Past. Staged Sound as Mediated Cultural Heritage* (Bielefeld: Transcript, 2013).

Birdsall, Carolyn, *Nazi Soundscapes: Sound, Technology and Urban Space in Germany* (Amsterdam: Amsterdam University Press, 2012).

Bocheński, Tadeusz, *Znam tylko Twoje listy* [I only know your letters] (Warsaw: Iskry, 1958).

Bołdok, Eng. Jarosław, [report], Polish Television Records Department, file no. A 85/2.

Czajkowski, Bogusław (ed.), *Ślady pamięci* [Traces of memories], radio broadcast, State Organisational Unit "Polish Radio and Television", 15 January 1988, Polish Radio Archives, file no. 18465.

F. P, "Zaczynamy 'Pierwszą Brygadą'" [We start with 'The First Brigade'], *Antena,* 30 (1935).

"Głos wolnej Warszawy. Inauguracja Polskiego Radia w stolicy" [The voice of free Warsaw: the inauguration of Polish Radio in the capital], *Życie Warszawy,* 13 February 1945.

Górski, Krzysztof, "Audycje religijne w Polskim Radiu" [Religious broadcasts on the Polish Radio], in Andrzej Ossibach-Budzyński (ed.), *Polskie Radio. Historia-program-technika. 90 lat Polskiego Radia* [Polish Radio: history, programmes and techniques. 90 years of Polish Radio] (Warsaw: Polskie Radio SA, ASPRA-JR, 2015).

Habielski, Rafał, *Polityczna historia mediów w Polsce w XX wieku* [A political history of the media in Poland during the twentieth century] (Warsaw: Wydawnictwa Akademickie i Profesjonalne, 2009).

Heller, Władysław, "Problem techniczny rozwoju radiofonii w latach powojennych" [Technical issues of the post-war development of radio broadcasting], Łódź, 15 August 1945, Polish Television Records Department, file no. A 132/12.

Krok, Edward, "Uwagi dotyczące radiofonizacji Państwa" [Notes on increasing state radio coverage], 15 May 1945, Polish Television Records Department, file no. A 85/14.

Kunert, Andrzej Krzysztof (ed.), *Mazurek Dąbrowskiego i inne pieśni narodowe* (Warsaw: Agencja Reklamowo-Wydawnicza A. Grzegorczyk, 2007).

Kwiatkowski, Maciej J., "Kalendarium Komitetu do spraw Radia i Telewizji 'Polskie Radio i Telewizja' za lata 1944–1973" [The radio and television committee "Polish Radio and Television": a timeline, 1944–1973], in The Collection of Maciej Kwiatkowski, Polish Television Records Department, file no. A 1852/3.

– *To już historia. Felietony o dziejach Polskiego Radia* [That's history now: columns on the history of Polish Radio] (Warsaw: Wydawnictwa Radia i Telewizji, 1975).

– *'Tu Polskie Radio Warszawa…'* [Polish Radio Warsaw here] (Warsaw: Państwowy Instytut Wydawniczy, 1980).

– "Jerzy Myśliński, Mikrofon i polityka. Z dziejów radiofonii polskiej 1944–1960" [Jerzy Myśliński: speaker and politician. From the history of Polish radio broadcasting 1944–1960], *Kwartalnik Historii Prasy Polskiej,* 31/1 (1990).

–, and Solecka, Krystyna (eds.), *Na warszawskiej fali* [On a Warsaw wave], radio broadcast, Radio and Television Committee, 13 February 1965, Polish Radio Archives, file no. 3979/1.

Markowska, Elżbieta, and Naliwajek–Mazurek, Katarzyna, *Okupacyjne losy muzyków. Warsaw 1939–1945* [The fate of musicians during the occupation of Warsaw in 1939–1945] (Warsaw: Towarzystwo imienia Witolda Lutosławskiego, 2014).

Michalski, Dariusz, "Radio poza radiem" [Radio beyond the radio], in Andrzej Ossibach-Budzyński (ed.), *Polskie Radio. Historia-program-technika. 90 lat Polskiego Radia* [Polish Radio: history, programmes and techniques. 90 years of Polish Radio] (Warsaw: Polskie Radio SA, ASPRA-JR, 2015).

Minkiewicz, Janusz, "Co słychać w Łodzi" [What you can hear in Łódź], *Przekrój,* 31 (1945).

Miszczak, Stanisław, *Historia radiofonii i telewizji w Polsce* [The history of radio and television in Poland] (Warsaw: Wydawnictwo Komunikacji i Łączności, 1972).

Nałkowska, Zofia, *Dzienniki 1945–1954* [Diaries 1945–1954], vol. 6, part 1 (1949–1948), ed. Hanna Kirchner (Warsaw: Czytelnik, 2000).

Nowicki, Robert, "Polskie Radio na Pradze" [Polish Radio in Praga], an interview with Jerzy Kotkiewicz, dir. Anna Mizikowska and Krzysztof Sagan, Nowicki Film-Production (7 September 2016), a documentary film produced in collaboration with the Praga Museum of Warsaw, a branch of Warsaw Museum, and the website *RadioPolska.pl,* https://www.youtube.com/watch?v=ckUbwJ6f6T0, accessed 10 September 2017.

Ossibach-Budzyński, Andrzej (ed.), *Polskie Radio. Historia-program-technika. 90 lat Polskiego Radia* [Polish Radio: history, programmes and techniques. 90 years of Polish Radio] (Warsaw: Polskie Radio SA, ASPRA-JR, 2015).

"Plan kablowej sieci transmisyjnej miasta Warszawy (26.06.1945r.), 'Schematy'" [Plan of the cable transmission network in Warsaw (26 June 1945), Schemata], no. 5, Polish Television Records Department, file no. A 24/1.

"Projekt budowy sieci rozdzielczej radiowęzła 'Praga'" [Design documentation of the "Praga" cable radio network], 15 June 1945, Polish Television Records Department, file no. A 131/7.

"Protokół III Walnego Zjazdu członków Związku Kompozytorów Polskich w dniach 20 i 21 października 1947r." [Protocol of the Third General Convention of the Polish Composers Union on 20 and 21 October 1947], ZKP/Polmic archives, file no. A 12/3.

"Raport z inspekcji w Dyrekcji Okręgowej P.R. w Warszawie przeprowadzony przez Insp. M. Orłowskiego w dn. 24–29 września" [Report from an inspection at the Regional Headquarters of Polish Radio in Warsaw performed by inspector M. Orłowski on 24–29 September], 3 September 1946, Polish Television Records Department, file no. A 4/3.

"Radio w zwierciadle prasy" [Radio in the mirror of the press], *Biuletyn Biura Studiów*, 3 (1946), 30, Polish Television Records Department, file no. A 1457/2.

Retmański, Leon, "Dot. radiofonizacji kraju" [On increasing radio coverage across the country], 7 August 1945, Polish Television Records Department, file no. A 85/19.

"Rozgłośnie, radiowęzły, radioabonenci" [Wireless stations, cable stations, radio subscribers], *Biuletyn Biura Studiów*, 2/5 (1946), file no. A1457/2.

Sagan, Krzysztof, "Jerzy Kotkiewicz o programie i sprzęcie na Targowej" [Jerzy Kotkiewicz on the programme and equipment at the radio station on Targowa Street in Warsaw] (1 February 2015), online cycle of recollections, in the series *Radio w pamięci zapisane* [Radio inscribed in our memories], http://radiopolska.pl/90lat/powojenna-odbudowa-polskiej-radiofonii/jerzy-kotkiewicz-o-programie-i-sprzecie-na-targowej, accessed 19 October 2017.

Sawicki, Tomasz, *Kronika roku jubileuszowego. 80 lat Polskiego Radia* [Anniversary chronicle: 80 years of Polish Radio] (Warsaw: Polskie Radio SA, 2005).

Semkowicz, Anna, *Rozgłośnia pracuje* [The station is working], radio broadcast, Radio and Television Committee, 12 March 1980, Polish Radio Archives, file no. PR III 77603.

Siezienewski, Andrzej, "Siezieniewski: Ważne w miejsce w życiu Szpilmana zajmowało Polskie Radio", http://www.polskieradio.pl/7/15/Artykul/446014,

Siezieniewski-wazne-miejsce-w-zyciu-Szpilmana-zajmowalo-Polskie-Radio, accessed January 2017.

– "Informacja i publicystyka" [News and journalism], in Andrzej Ossibach-Budzyński (ed.), *Polskie Radio. Historia–program–technika. 90 lat Polskiego Radia* [Polish Radio: History, programming and technology. 90 years of Polish Radio] (Warsaw: Polskie Radio SA; Oficyna Wydawnicza Aspra-JR, 2015).

Sommer, B., "Sprawozdanie Wydziału Eksploatacyjnego Dyrekcji Technicznej Polskiego Radia na miesiąc lipiec 1945 roku i plan pracy Wydziału Eksploatacji Dyrekcji Technicznej P.R." [Report by the Maintenance Department of the Technical Directorate of Polish Radio for June 1945 and work plan for the Technical Maintenance Division of the Polish Radio Technical Directorate], 23 July 1945, file no. A 85/19.

Spotkania z Warszawą [A meeting with Warsaw], radio broadcast, Polish Radio Central Station in Warsaw, 16 January 1959, National Digital Archives, file no. 33-T-1248.

"Sprawozdanie z działalności Polskiego Radia za wrzesień i październik 1945r." [Report on the activities of Polish Radio in September and October 1945], Polish Television Records Department, file no. A 85/11.

"Stan czynnych radiowęzłów Polskiego Radia na dzień 25.05.1945 r." [Status of active Polish Radio cable stations on 25 May 1945], Polish Television Records Department, file no. A 132/12.

"Statystyka programowa" [Broadcast statistics], *Biuletyn Biura Studiów*, 3 (1945), Polish Television Records Department, file no. A 1457/2.

Sterne, Jonathan, *The Audible Past: Cultural Origins of Sound Reproduction* (Durham: Duke University Press, 2003).

Strycka, *Wspomnienia radiowców* [Memories of radio producers], radio broadcast, 1972, Polish Radio Archives, file no. ARCH. 10428/1.

Szelegejd, A., "Plan prac na m-c grudzień 1945r. Działu Montażowego Wydziału Budowy Radiowęzłów" [Work plan of the Installation Division of the Cable Radio Building Department for December 1945], December 7, 1945, Polish Television Records Department, file no. A 85/19.

Szpilman, Władysław, *Pianista* [The pianist] (Cracow: Znak, 2002).

Tomaszewski, W., "Plany prac działu sieciowego wydz. TRB" [Work plan of the network division of the Cable Radio Building Department for December], 7 December 1945, Polish Television Records Department, file no. A 85/19.

"Techniczny i finansowy plan radiofonizacji przewodowej kraju na rok 1946" [Technical and financial plans for increasing national radio coverage for 1946], Polish Television Records Department, file no. A 132/12.

Wilkowicz, Wiktor, "Techniczny i finansowy plan przewodowej radiofonizacji kraju na rok 1945" [Technical and financial plans for increasing national radio coverage for 1945], 14 June 1945, file no. A 132/12.

– "Sprawozdanie stanu robót Wydziału TRB na dzień 26.05.1945 r." [Status report of the Cable Radio Building Department on 26 May 1945], 28 May 1945, file no. A 85/19.

– "Sprawozdanie według sprawozdań poszczególnych Dyrekcji Okręgowych" [Report based on reports from Regional Directorates], 25 October 1945, file no. A 132/12.

Witort, A., "Plan prac Wydziału Budowy Radiowęzłów na m-c listopad 1945r." [Work plan of the Cable Radio Building Department for November 1945], November 9, 1945, Polish Television Records Department, file no. A 85/19.

"Wykaz Dyrekcji Okręgowych PR wg ilości radiowęzłów, abonentów głośnikowych i długości linii" [Register of Polish Radio Regional Directorates by number of cable radio stations, loudspeaker subscribers and network lengths], 27 April 1946, file no. A 132/12.

"Wyniki ankiety programowej" [Results of the programming survey], *Biuletyn Biura Studiów,* 2/5 (1946), Polish Television Records Department, file no. A 1457/2.

Zadrowski, Witold, *25 lat Polskiego Radia* [25 years of Polish Radio], radio broadcast, Polish Radio, 15 July 1969, Polish Radio Archives, Literary Broadcasts Department, file no. C/9651, 6107/2.

"Zarządzenie Dyrektora Naczelnego W. Billiga nr 24" [Decision no. 24 by Head Director W. Billig], 28 April 1945, Polish Television Records Department, file no. A 13/1.

"Z frontu radiofonizacji kraju" [From the frontline of national radio broadcasting expansion], *Biuletyn Prasowy Polskiego Radia, Wydział Prasy i Propagandy,* 6/14 (1946), Polish Television Records Department, file no. A 1457/1.

Zoriana Rybchynska

Ukrainian Catholic University, Lviv

The voices of a liberated/occupied city: The Lviv soundscape of 1944–1946 in Ryszard Gansiniec's journal

Abstract: The daily notes made by Prof. Ryszard Gansiniec from 1944 to 1946 are unique statements of a Polish citizen of the city, who lived under the Soviet administration and experienced daily changes in the urban landscape. Gansiniec was a professor of Classics at John Casimir University in Lviv (1919–1939, 1944–1946). He remained in Lviv during the Second World War, but he was forced into exile by the Soviet regime and had to leave Lviv in 1946. His notes contain detailed descriptions of the changes in the city and hostile comments addressed to the new regime. He believed that Lviv would remain a Polish city. In fact, his Polish nationalist ideology, classical education and personal qualities, such as an uncompromising nature, filter the image of the city presented in his journal. These descriptions of the city give us an opportunity to reconstruct the audiosphere of the city and analyse how sensitivity to sounds changed in the new culture.

Keywords: city soundscape, auditory topoi, own – other, keynote sounds, sound marks, sonic icons, Lviv

The historical fate of Eastern Europe during the twentieth century was marked by the domination of nationalistic ideology and totalitarian regimes, the tragic events of two world wars and two genocides (the Holocaust and the Holodomor), mass resettlement, and a sea change in the cultural landscape, deprived of the variety and polyphony developed over the centuries. Reflection on that difficult history has become an important part of the process of reconstructing the national identity of countries in the region that are experiencing political, economic and social transformations. Within this context, the case of Lviv – a city on the cultural border, now a Ukrainian city with a complex and not entirely accepted history – attracts the attention of historians,[1] who, especially in recent years, have proposed a more complex and controversial picture of the relationship between the various ideological, national and cultural projects implemented within the space of the city. In addition to works that build up an overall picture of events and

---

1 e.g. Snyder, *Bloodlands.*

processes, no less important is attention to the fate of individuals, to biographical sources and evidence.[2] One such interesting and at the same time little-known source consists of the notes made in 1944–1946 by Professor Ryszard Gansiniec, written as letters to his wife.[3] They now constitute a valuable resource, providing an eyewitness account of the changes which the inhabitants of Lviv experienced during the dramatic times of the post-war division of Europe and the establishment of new national borders. Let us look at them more closely.

One late June evening in 1944, Gansiniec is sitting in his cool Lviv flat at a gas lamp and writing to his young wife Zofia, who has left the city with their children to escape the trials of war. It's quiet. His eyes hurt because of the flickering light, and his fingers are numb from the cold. But the Professor describes his day at the university carefully and accurately: whom he met, what they said, what's going on in the city. Despite his bitterness at the changes taking place in Lviv, with the strengthening of the Soviet regime, he stoically clings to the hope that everything will turn around and the city will return to the "homeland". That is why he has stayed behind ("As long as I'm in Lviv, it is a Polish city", he says on one of the first pages). He is still teaching at the university and trying to protect his library. Bitterly mistaken in his expectations, he is steadfast in his resistance to the changes, which he observes not only at the university, but also in the city's streets, markets and shops. Although not born here, Lviv has become to him, a researcher of ancient culture, a symbol of his beliefs and faith, formed during the Second Polish Republic, at a time when the ideology of nationalism was prevailing.

Ryszard Gansiniec was born in 1888 in Upper Silesia. He developed his beliefs while studying Classics and German at the universities of Münster and Berlin, and in the Prussian army. In 1919, two years after obtaining his doctorate, Gansiniec became Head of Classics at the University of Lviv, where he conducted numerous research projects and produced his publications. In 1946, after experiencing a complete change in the structure of his life, with a heavy heart, he left the city and went to Poland. This resolute man, who held firm to his beliefs and rejected all compromise, was destined to go through a period of serious political, social and cultural transformations changing the face of the whole of Europe, affecting not just the fight against the great powers and the struggle for nationhood, but also the life of cities and their inhabitants, cast into the vortex of political games. Unlike many other cities in this part of Europe, Lviv seems to have been lucky, because the wartime destruction was minimal. At the end of July 1944, three years

---

2 One of the best works of recent years is Ola Hnatiuk's *Odwaga i strach.*

3 Gansiniec, *Notatki Lwowskie.*

after the Nazi occupation, in the dead silence of the early morning, the Soviet army entered the city.[4] This army was no longer greeted in the streets as it had been in 1939, at the first entry of Soviet power, resulting from the Molotov-Ribbentrop Pact. The citizens of Lviv no longer had any illusions about this regime. The majority expected that the fate of the city had not been determined and that it could not remain within the borders of the Soviet Union. Meanwhile, despite serious political changes, daily life continued. The gradual formation of a new mode of urban life can be reconstructed mainly on the basis of the notes written by Ryszard Gansiniec.

This reconstruction will specifically focus on the peculiarities of the soundscape that in recent years has been the centre of interest among researchers from different fields: historians, anthropologists, urbanists, cultural theorists and music scholars. As is rightly observed by Mark M. Smith in his introduction to the reader *Hearing History*:[5] "That question of sound, noise, and aurality will not just infiltrate historical narratives but also change the very conceptualization of historical thinking". Sound studies are now offering new ways to look not only at the past, but also at a city's cultural landscape, and at different representations of cultural, national and political identity, etc. Yet we cannot simply listen to our urban past. We encounter a rich cultural heritage of city sounds presented in texts (e.g. memoirs or journals), radio broadcasts and films. Specifically, this approach is presented in the collection produced by members of the Soundscapes of the Urban Past research project at Maastricht University. According to Karin Bijsterveld, "our knowledge of past soundscapes, transient and intangible as they are, is therefore largely dependent on historical texts in which people described what they heard and what these sounds meant to them".[6] Research into such sounds feeds our imagination, because we often reconstruct the soundscape of a city also on the basis of visual archive documents, descriptions and images in newspapers or memoirs. This all gives us "a unique chance to study the dramatization of urban sound over time, and thus to understand the varying and changing representations of urban identities".[7]

---

4 From the memories of Jadwiga Jamrozhowna, "Horses of the Great War": "Kulparkowska Street […]. No one greeted them. The shuttered windows of the neat suburban houses testified not only to the early hour. We already knew what awaited us under the rule of the victors. There was no place to hide from this nightmare". Quoted in Fastnacht-Stupinicka, *Zostali we Lwowie*, 76.

5 Smith, "Onward to audible pasts", XXI.

6 Bijsterveld. Introduction, 14.

7 Ibid.

"The historical actors – as Karin Bijsterveld further explains – articulate their perception of sounds only when particular sounds moved them, in either a positive or a negative sense".[8] That is why the description of sounds is often used to convey a certain atmosphere or characteristics of a particular community or culture. Bijsterveld says that in trying to explain the meaning of sounds and "thereby rhetorically strengthening their position, authors of the descriptions employed particular repertoires of dramatizing sound".[9] Among the repertoires, Bijsterveld distinguishes specific auditory topoi, such as intrusive sound, sensational sound, comforting sound and sinister sound.[10] She proposes to use them as a key tool for the study of soundscape changes and its representation in historical texts, especially for determining the author's strategy for presenting an urban landscape in which different cultures collide. In addition, as Bijsterveld suggests, these texts can be explored with the use of such concepts proposed by Raymond Murray Schafer as "keynote sounds", "sound marks" and "sonic icons".[11] This approach will distinguish the types of sounds depending on their functions in the creation of a city soundscape.

Already on the first pages of Gansiniec's notes, we find characteristic auditory topoi that are attributes of those times. The memory of peace (i.e. silence) in June 1944 is treated as a sign of the return of a peaceful atmosphere: "Here in Lviv it is quiet [...] and since you left there has been no alarm".[12] By December, however, almost every entry records the current audiotopos of horror and fear ("sinister sound"), defining the atmosphere of the city, the release from which actually brought re-occupation by another totalitarian regime. The author captures characteristic sounds reinforcing the omnipresent fear and anxiety, which are associated with emptiness and darkness, with the violence and crime that are dominant on the streets:

> 16. XII. 1944. A separate chapter is our Lviv evenings in general. It's already dark by six, the soldiers prowl the roads and snatch watches, handbags, coats, wallets, flashlights – whatever they can. The screams, the cries... Sometimes shooting and rattling. [...]. So

---

8 Ibid.

9 Ibid., 15.

10 Karin Bijsterveld borrowed the concept of auditory topoi from the works of Michael Cowan and Philipp Schweighauser on the textualisation of sound and noise in poetry and literature: Cowan, "Imagining modernity"; Schweighauser, *The Noises*.

11 Schafer, *The Soundscape*. The last notion was introduced by Hans-Jürgen Krug in *Kleine Geschichte des Hörspiels*.

12 Gansiniec, *Notatki*, 6.

> our street is empty after six, and anyone who has to go out runs back home to be under their own roof again.[13]

The atmosphere of general oppression disturbs the only hope for salvation, paradoxically associated with the continuation of the war. That is why Richard Gansiniec so vigilantly captures the emergence of new sounds on the streets, caused by the greater movement of military equipment in the spring of 1946:

> 16. III. 1946. It is 11 p.m. The columns of cars and troops are moving non-stop down Pelczynska Str., the military are singing "Sviashchennaya voyna" [holy war]. Here in Lviv, things are getting lively: tanks rattling along the cobbled streets, military trucks I've never seen before hurtling along, marching troops. There's a car in front of a building where the Soviets are loading their belongings and their brats to flee Lviv in time. The feverish panic is subsiding; people are beginning to reason, speaking carefully and stocking goods.[14]

Only his private flat seems to be a haven of peace and quiet, where he can still focus on his work. Other places (cathedral and library) are already subject to stringent controls:

> 8. XI. 1945. I was thinking of working at the Academy, because it was quiet there and there was a good library, so I wanted to rewrite the work on a typewriter. Yet when I was the only one sitting there, the manager came and said that it had to be closed down and sealed up the typewriter. I had to get out.[15]

Finally, private space does not guarantee shelter, and city dwellers are exposed to physical or phonic interventions from newcomers:

> 16. XII. 1944. If I can, I avoid being outside in the evening. Although they come here as well, and even at night. For example, today three of them came and wanted to search. I didn't let them in, although they hit the door with their boots and rifles and made a noise.
>
> 8. XI. 1945. Naturally they're still knocking on my door under various pretexts, but I don't speak. Once again they're belting out love songs, dancing and drinking, so my work doesn't get done.[16]

In this situation, the only natural sounds which the author notes are the sounds of wind, strengthening the sense of vulnerability and helplessness:

> 13. XII. 1945. The wind and blizzard are carousing outside; here at home there's a real concert going on with the whistling.[17]

---

13 Ibid., 14.
14 Ibid., 134, 139.
15 Ibid., 48.
16 Ibid., 14, 48.
17 Ibid., 63.

> 18. II. 1946. The wind rumbles, whimpers, groans and roars – the gaslight is tiny and still goes out in the wind.[18]

> 3. III. 1946. Outside, the cold, frosty storm is howling like a pack of Russians.[19]

In this series of descriptions, it is easy to see how sound impressions acquire additional cultural meanings and political contexts. The comparison of a storm to a pack of Russians in particular is highly characteristic of the whole text, because the world of post-war Lviv is constructed here on the basis of the anthropological opposition between "own" and "other". Such a structure is the result not just of individual characteristics but mainly of the radicalisation in the world during and after the war, where the underlying determinant of the person, space and culture is again the primary division between own and other. From this unambiguous and harsh perspective, Ryszard Gansiniec perceives all the events and all the people, makes conclusions and evaluates by using in his descriptions different types of sounds, which are referenced to meaningful axiological semantic complexes.

In his descriptions of everyday events, the author often mentions sounds forming an audio background that impresses on the listener what is happening around him. In the short sketches depicting Lviv's streets, Gansiniec mentions screams, drunken quarrels and the indifferent silence of observers:

> 8. XI. 1945. Today I was in the city for dinner. It's not nice to walk on the streets: Russian culture is everywhere; people vomit in the middle of the pavement and relieve themselves against the walls. Lviv, as someone said, is quite messy [*zafajdany*]. There are a lot of drunks, antagonising each other.[20]

> 10. XI. 1945. Sometimes drunks fight in broad daylight. On Sophia Square, someone who had been trampled in the face was lying in blood; the officers looked at it, but said nothing.[21]

Another constant theme is the everyday queuing in shops, where Gansiniec records sound signs expressing the aggressivity and hostility associated with those places, which he perceives as characteristic of the culture of newcomers to the city:

> 1. VI. 1946. A new month. Yesterday was an unlucky day for me – the last day of the month. You have to redeem vouchers for flour and other products, but the products weren't given out, and only at 1 p.m. were sales permitted. Imagine what happened: queues formed in a large warehouse, and a hubbub of shouting, disputes and fights erupted for a place in the queue.[22]

---

18 Ibid., 112.
19 Ibid., 125.
20 Ibid., 48.
21 Ibid., 49.
22 Ibid., 186.

The author tries to oppose both the alien and angry crowd and also the new order, so as to reach his goal "with dignity", for instance thanks to the protection of Polish women.

> 29. IV. 1946. Can you believe that from 9 till 7 I queued in Gastronom [grocery store] to choose products before the end of the month? I was helped by Polish women, otherwise I wouldn't have got anything. The "Sovietki" are unruly and specialise in queuing – the screaming and yelling are unheard of and constant. I always stay aloof and avoid quarrelling, so finally I also arrive at the end, more worthy.[23]

In this passage, conflict between "own" and "other" is indicated, and it is visualised by featuring the characters of Polish women and "Sovietki" (women who came to Lviv after the war from eastern Ukraine or Russia). For Professor Gansiniec, these Soviet women personify the key features of Soviet culture:

> 3. II. 1946. Every Sovietka can be identified by her pretentious elegance, contrasting with her constitution. They all have brightly painted lips, but not all of them have money for powder. These "ladies" also have their nails painted bright red, but not necessarily clean. They talk loudly, and their voices don't have the softness and rich tones of our women; they only modulate the voltage, speaking more loudly or more quietly; in general, the tendency to scream is considered quite elegant. In addition, their laughter is loud, not modulated, and produced by the whole face, as if it were nothing but massive face muscles. Their tastes are quite primitive [...]. Their whole idea of taste is just to accumulate and show off their wealth, as well as their merriment: loud screaming and commotion, celebration, eating and drinking themselves unconsciousness, listening to the gramophone and roaring.[24]

This entry contains distinct sonic icons, phonic features of great importance, which the author uses to create a comprehensive picture of a stranger. The sonic icons are closely related to visual characteristics (a woman's dress and make-up, even physiological peculiarities, like the musculature of the face), and they are also based on the opposition between "own" and "other" (a Polish woman and a Soviet woman).

By describing the sonic environment of the public space of Lviv, the author creates another contrast, just as powerful. It is clear how the part of the city subjected to the harsh control of the new authorities expanded during the years 1944–1946, while the "own" space decreased catastrophically. Diligently recording these processes, Gansiniec seriously considers the "spiritual exercise" of Soviet citizens and gives an exact analysis of the Soviet propaganda machine, in which phonic control

23 Ibid., 166.
24 Ibid., 100.

and pressure play a very important role, placing a person under the constant influence of a sonic environment constructed by radio, meetings and agitators. This "taming" leaves "Soviet Russians" only one space for freedom, where the lowest human instincts are located ("Their desire heads in a different direction: they must stuff themselves, get drunk, steal and hurt each other"[25]). They express their instincts in an incontinent and aggressive phonic form, in a "fever binge": "it was the only remaining outlet, like vodka was the only item without a queue".[26] This aspect of Soviet culture is treated by modern scholars as a manifestation of archaic ways of experiencing the vitality of an orgiastic immersion in bodily sensations (eating, drinking) and the intemperate expression of emotions, as compensation for the social coercion and political pressure in everyday life, characteristic of totalitarian cultures.[27]

> 12. II. 1946. Here in Lviv, already at 7.30 in the morning the police have driven people to the election and checked every house till late in the evening to make sure everyone went. They didn't come to me. It was a point of honour for every patriot to drink himself into a stupor and sing wholeheartedly.[28]

The Classics professor treats these symptoms within the context of the clash of civilisations – Europe and the Steppe, East and West – that was familiar in Polish Romantic literature and was revived in the nationalist discourse during the Second Polish Republic. He even notes the "bestial nature" of the newcomers from the East during New Year's celebrations: "31.12.1945. I'm waiting for midnight to greet the New Year. There is shooting from rifles, revolvers and automatic firearms, and all the mouths let out a roar, like steppe wolves – thus they welcome the New Year".[29] The entire cycle of the Soviet calendar is celebrated in the same atmosphere of "taming and terror". What remain unchanged are the mass scale and the coercion (meetings, demonstrations, checklists of participants, etc.) and the high-volume sounds ("1. V. 1946. Cannon shots and colourful rockets: everything trembled, but not one windowpane flew out"; "9. V. 1946. Today again at 10 in the evening there was heavy artillery, searchlights, colourful rockets and the firing of machine guns and revolvers"[30]). A separate place is occupied by the

25 Ibid., 111.
26 Ibid., 128.
27 e.g. Malte. *Soviet Mass Festivals*.
28 Gansiniec, *Notatki*, 107.
29 Ibid., 74.
30 Ibid., 167, 171.

first Soviet elections, prepared as an important political campaign of an explicitly festival-like character:

> 5. I. 1946. The whole city again has the appearance of a public manifestation: there are portraits, notices, posters and leaflets everywhere, all arranged for the election on 10 February. The choirs are practising, the orchestras are playing, the schoolchildren are singing in the streets. There is such a great commotion for the election, in which you have to choose between Stalin and Dzhugashvili.[31]

In opposition to the space dominated by the Soviet regime, there is another public space described in Gansiniec's notes with which he identifies himself. This is the pre-war and pre-Soviet sonic environment, characterised by sounds associated with churches (bells and masses). Although beyond the control of the new power, this sonic environment is the object of constant attacks. The author represents this environment for the sake of diversity and to accentuate the contrasting characteristics: singing instead of screams, bells instead of shots, restraint instead of aggression, spirituality instead of carnality, reconciliation rather than hostility.

> 25. XII. 1944. Today is a holiday. Yesterday at five o'clock I was at the Christmas Mass in St Magdalene's. The church was crammed, and all the Soviet women and girls were also there. We sang carols. Those traditional Polish songs are beautiful – it was nice to hear.[32]

The impact of this sacred space, through religious celebrations, at some point expands, for instance, to the halls of the university. This dissolves the author's hostility towards the Ukrainians. This kind of reconciliation happens at Christmas time:

> 25. XII. 1945. Today is a working day like any day – only the Ukrainian world around me was still in a Christmas mood, because I receive wishes wherever I go. When I came to the first year students' class, I was welcomed with a long Latin greeting, then the students sang three songs for me: "Christus, Christus Natus est nobis"; the Polish Christmas song "Gdy się Chrystus rodzi" and finally a Ukrainian carol. The whole class is Ukrainian and the university (on St Nicholas Day) is Soviet, so people kept looking through the door to see what it was all about. […] From the street come carols sung by children.[33]

In one of the final notes, from 2 June 1946, the author writes about the last mass in the churches as a final testimony to the capitulation, related to the decision to leave Lviv at the time of the forced resettlement of Polish and Ukrainian people as a result of the re-drawn boundaries of post-war Europe.

---

31 Ibid., 82.

32 Ibid., 17.

33 Ibid., 69.

In a sense, Gansiniec's notes represent a reliable document of the loss of the city, which was a key symbol of national identity in the professor's inner world. This predominant emotional climate of resistance and defeat greatly impacts on the author's narrative strategy, which was created on the basis of the conflict between "own" and "other". The distinctive repertoire of dramatising sounds is subordinated to this strategy and consists of auditory topoi, created from intrusive, sensational and sinister sounds. In Gansiniec's presentation of the Lviv sonic environment of 1944–1946, the sounds perform various functions: they create an emotional background to the described events and characters; they are characterised by certain places; they mark "own" and "other" space, participating in the creation of symbolic figures and images of the different cultures.

In addition to this, the professor pays attention to the sonic features of the culture of the newcomers (such as insistent noise, phonic incontinence and high levels of volume). His considerations remarkably illustrate Theodor Adorno's idea of the political nature of sound, according to which silence, music and noise are cultural constructs and are used to create and strengthen the social system.[34] Especially in the case of totalitarian systems, the sonic environment is an instrument of control and social pressure, as exemplified by the presence of loudspeakers on the Soviet streets, constantly broadcasting the leaders' speeches and turgid music, and the later appearance in Soviet culture of the image of "big ears" eavesdropping on citizens' lives.

So the receiver and recorder of sounds is also a necessary participant in the production of the sonic environment. Specifically, listeners impart different meanings to sounds, according to their higher or lower phonic sensibility, the type and level of their education, the ideological system, etc. From his notes, Prof. Gansiniec comes across as a very sensitive listener, a proponent of nationalistic ideology, and well-educated in the Classics. This last feature allows us to find very interesting comments about the presence of the Polish language on the streets, as well as attempts to reproduce the sound of Soviet newspeak and highly characteristic sound metaphors. One of them, at the very beginning of the text, expresses hope: "10. XII. 1944. […] Like most people, I would like to live through the winter sleeping like a bear, and at the sound of Easter bells wipe my sleepy eyes and hear the voice that has finally stopped the clanking of war".[35] And his last entry describes his preparations for leaving the city: "7. V. 1946. Today I was packing my books. You know, it's very tiring and very sad work – almost as if a dead man was being put

34 Adorno, "On the social situation", 391–436.

35 Gansiniec, *Notatki,* 11.

in a coffin, and when the carpenter hammers down the lid it sounds just like the nailing of a coffin in the gloom of the Dominican coach house. So I'm sad, very sad. I feel that I have to say goodbye to something close to me and very dear".[36] Reconstructed from Gansiniec's notes, the soundscape of Lviv from 1944 to 1946 is not of course complete and objective. However, its value lies in its representation of the complex and ambiguous history of the cities of Eastern Europe, in the discovery of the polyphony of national, political and cultural identity, and in demonstrating the cognitive value of research into the sonic environment of a city.

Bibliography

Adorno, Theodor, "On the social situation of music", in *Essays on Music* (Berkeley: University of California Press, 2002), 391–436.

Amar, Tarik, *The Paradox of Ukrainian Lviv. A Borderland City between Nazis, Stalinists, and Nationalists* (New York: Cornell University Press, 2015).

Bijsterveld, Karin, Introduction to Bijsterveld (eds.), *Soundscapes of the Urban Past. Staged Sound as Mediated Cultural Heritage* (Bielefeld: Transcript, 2013), 11–28.

Cowan, Michael, "Imagining modernity through the ear", *Arcadia*, 41 (2006), 124–46.

Fastnacht-Stupinicka, Anna, *Zostali we Lwowie* [They stayed in Lviv] (Wroclaw: Sator media, 2010).

Gansiniec, Ryszard, *Notatki Lwowskie (1944–1946)* [Lviv notes (1944–1946)] (Wroclaw: Sudety, 1995).

Hnatiuk, Ola, *Odwaga i strach* [Courage and fear] (Warsaw: Wydawnictwo KEW, 2016).

Krug, Hans-Jürgen, *Kleine Geschichte des Hörspiels* (Konstanz: UVK Verlagsgesellschaft, 2003).

Malte, Rolf, *Soviet Mass Festivals, 1917–1991* (Pittsburgh: University of Pittsburgh Press, 2013).

Mick, Christoph, *Lemberg, Lwów, L'viv, 1914–1947: Violence and Ethnicity in a Contested City* (West Lafayette: Purdue University Press, 2015).

Schafer, Raymond Murray, *The Soundscape. Our Sonic Environment and the Tuning of the World* (Merrimac: Destiny Books, 1993).

Schweighauser, Philipp, *The Noises of American Literature, 1890–1985. Toward a History of Literary Acoustics* (Gainesville FL: University Press of Florida, 2006).

36 Ibid.

Smith, Mark M., "Onward to audible pasts", Introduction to Smith (ed.), *Hearing History. A Reader* (Athens: University of Georgia Press, 2004).

Snyder, Timothy, *Bloodlands: Europe Between Hitler and Stalin* (New York: Basic Books, 2012).

Karolina Jara

Institute of Art History, University of Wrocław

The soundscape of public space in Breslau during the period of National Socialism

Abstract: This article examines the methods of space appropriation employed by the Nazi authorities, with particular emphasis on the musical and auditory layers and the projects for reshaping the public space in Breslau, focusing on the designed sound environment. The first of the issues presented, the 12th German Choral Society Festival (Deutsches Sängerbundesfest), took place from 28 July to 1 August 1937. It was the biggest musical event organised in Breslau under the Third Reich and one of the most important events on the national level. The festival allows us to trace how the musical sphere was harnessed in the space of the city and how its propaganda potential was exploited. The second phenomenon under scrutiny concerns a space that was designed but never actually existed, so the question of its acoustic characteristics seems completely inaccessible. Useful in this respect are architectural plans and photographs of mock-ups with planned areas of the city that allow us to reflect on designed architecture in relation to the planned soundscape of new urban spaces. I am most interested in how and to what extent sound can be considered an important factor in the process of redesigning an entire city centre in the late 1930s. The aim of this article is to broaden understanding of the sound experience of the Nazi city, as well as its role in urban planning.

Keywords: public space, architecture, Third Reich, music festival, space appropriation, designing the sound environment

> The air is exceptionally warm tonight. The strains of a brass band are wafted from far away, so it seems, by a gentle breeze. The sound fluctuates, reverberating from distant streets. The music grows louder. With a sudden explosion of noise, the band merges from a side street followed by a detachment of Brownshirts marching in step. They, in turn, are followed by a group in regional costume and some civilians who have joined the procession en route. Many windows are open, and the din invades the rooms behind the fluttering curtains. Across the way, curious residents are already leaning on their window sills and gazing down into the street. Many wave. The window of one darkened room is closed and the curtain drawn as if by some ghostly hand. The windowpane at my back begins to vibrate. The night air resounds to the blare of trumpets and the rattle of snare drums. They're passing the house now. The color bearer's flag, propelled by a headwind, slaps him in the face.[1]

1 Beyer, *The Karnau Tapes*, 67.

Given the abundance of elements that create the sound environment of a particular place, it is essential to clarify the areas that we look at, or rather "listen to". The soundscape in Breslau under the Third Reich may be analysed from various angles, such as the soundscape of the small number of newly erected public buildings, the functioning of musical institutions or radio broadcasts, and this study looks at only a small part of this very complex phenomenon. Based on the example of a music festival, it shows the methods of space appropriation employed by the Nazi authorities, with particular emphasis on music and sound. It also describes the interaction between the public space and the architecture that forms it, and also their acoustic qualities, within the context of the plans for the thorough redesigning of the city centre.

The 12th German Choral Society Festival (*12. Deutsches Sängerbundesfest*) took place from 28 July to 1 August 1937 and was the most important musical event organised in Breslau under the Third Reich and one of the most important events on a national level. It allows us to trace how the musical sphere was incorporated into the space of the city and how its propaganda potential was exploited. Observations regarding the auditory impact of the space itself are preceded by remarks on the arrangements for the festival, in particular those related to architectural works and temporary decorations. This is important to the subject, because the urban fabric, as a material and visual factor, is intrinsically linked to the soundscape of a place. Moreover, this planned spectacle for a mass audience combined visual and auditory elements in an inseparable whole. The analysis of the festival also covers phenomena related to the history and musical culture of Silesia, which formed a symbolic layer of the event.

The second phenomenon under scrutiny concerns a space that was designed but never actually existed, so the question of its sound characteristics seems completely inaccessible. Useful in this respect are architectural designs and photographs of mock-ups with planned areas of the city that allow us to reflect on the designed architecture in relation to the planned sound environment of new urban spaces.

What interests me most is how and to what extent sound can be considered an important factor in the process of redesigning the entire city centre during the late 1930s. Did the existing urban soundscape influence the new plans? What does the case study of Breslau allow us to say about the designing of city acoustics under the Third Reich? I will attempt to broaden understanding of the sound experience of the Nazi city, as well as its role in urban planning.

The quote from Marcel Beyer's novel *The Karnau Tapes* at the beginning of this article is a literary representation of part of the German soundscape under the Third Reich. Not only is the plot set in Nazi times, the book is also devoted to sound, as the main character, Hermann Karnau, works as a sound technician and is obsessed with sound and with recording the human voice.

Arrangements

Grüß Gott! Grüß Gott mit hellem Klang! Heil deutschem Wort und Sang!
Singers' greeting (*Sängergruß*)[2]

The German Choral Society (*Deutscher Sängerbund*) was founded in 1862 to unite the numerous singing organisations that flourished during the first half of the nineteenth century and organised rallies of male choirs.[3] Apart from the concerts and rivalry, festivals also manifested national ideas that were gaining strength in the tense political arena before the unification of Germany in 1871.

This was particularly evident at festivals held in border cities, hence the choice of Breslau. Press and radio coverage used mostly the characteristic form of narration that I call "border discourse". This was present in the official statements of the Nazi authorities, and it intensified during the late 1930s, before the outbreak of war. It highlighted the difficult situation of the borderland territories of the Reich, especially the disputed areas that were nationally, ethnically, religiously and culturally diverse and therefore susceptible to "foreign" influences. Such discourse was designed to create a sense of threat and the need to strengthen the "German character" of those areas. So holding a nationwide German festival in Silesia was also significant in terms of propaganda. The Breslau festival was the first and the last to be organised under the National Socialist regime. The previous festival took place in 1932 in Frankfurt am Main; the next edition was planned for the 1940s in Vienna, but it did not take place because of the war.

It should be stressed that this was not the first time Breslau was chosen as the venue for a choral gathering lasting several days. The previous festival in the city was held exactly thirty years earlier, in 1907, in a temporary purpose-built hall in the park on the Friebeberg (now between Sztabowa Street and Powstańców Śląskich Square),[4] in a remote southern part of the city. The 1937 festival took place in partly new facilities. That was a big asset of the city and, apart from geopolitical issues, it certainly contributed to Breslau being chosen again. The main venues were the Jahrhunderthalle (now Hala Stulecia) with adjacent exhibition grounds, opened in 1913, and the sports complex with stadium built in the 1920s in Leerbeuthel (now the Olympic Stadium in Zalesie). Yet the presence of a large

2 Loeser and Werbeck, *Musikfeste im Ostseeraum*, 13.
3 Fastl, "Sängerfest".
4 *Schlesische Tageszeitung*, 27 June 1937.

number of choirs also necessitated the construction of temporary stages. A sheltered podium was created at the stadium to accommodate 40,000 choral singers (Fig. 1), and a special stage with steps in the Jahrhunderthalle, both providing adequate acoustics and visibility (Fig. 2). Breslau also had numerous concert halls, including spaces of unique historical and artistic value, such as the Aula Leopoldina and Oratorium Marianum.

Figure 1: Chorfeier *[Choral celebration], reprod. from* Deutsches Volk – Singend Volk. Herausgegeben im Einvernehmen mit dem Deutschen Sängerbund *(Berlin: C. A. Weller, 1937), 79. Courtesy of Wroclaw University Library*

Figure 2: Noch einmal: Feierliche Eröffnung *[Once again: grand opening], ibid., 21*

In addition to the venues that enabled concerts and shows to be organised, musical infrastructure was also important. The Jahrhunderthalle had an organ made by the Wilhelm Sauer Organ Company, which at that time was one of the largest in the world. For the purpose of the festival, the number of registers was extended from 200 to 222, under the masterly supervision of the organist Gerhard Zeggert.[5]

Richard Konwiarz, an architect and municipal building officer, was responsible for the image of the city projected at the Sängerbundfest. Konwiarz, associated with Breslau from 1909, was a close associate of Max Berg and was involved in Berg's Jahrhunderthalle project. Konwiarz later designed the stadium in Leerbeuthel and many other important buildings in Breslau, up to 1945.

Apart from the actions taken by the state and municipal authorities, the preparations also involved the residents, who received guidelines on how to adorn their houses. In addition to the general remarks that "decorations [...] should be in the best taste, and kitsch is undesirable", permissible forms and colours of ornaments were also specified, primarily including flags, garlands and fresh green wreaths.[6]

5 *Breslauer Gemeindeblatt*, 21 July 1937, 234.

6 Ibid., 233.

The choice of Breslau as the singing festival venue also provided an opportunity to familiarise participants with Silesian musical culture and present individuals who had contributed significantly to its development and were not opposed to the state ideology.

The historical patron of the festival was Henryk IV Probus a thirteenth-century prince of the Piast dynasty.[7] For the festival, a temporary statue of the prince, made by Viktor Eichler, took pride of place in the heart of the city (Fig. 3). The plaster statue, covered with bronze colour paint, was placed on a pedestal on the east side of the Market Square. The emphasis was on the authenticity of the prince's image, which was modelled on the effigy on his tomb in the Church of the Holy Cross and a drawing from the Codex Manesse, the most important medieval collection of German poetry. The prince was depicted as a singer and a soldier. His talents as a Minnesinger and his merits on the battlefield were reflected in the violin in his hand and the shield and sword at his feet, and also in the inscriptions on the pedestal, including a passage from one of the two pieces of Minnelied attributed to him.

7 *Schlesische Tageszeitung*, 29 June 1937.

Figure 3: Herzog Heinrich IV, Schlesiens Minnesänger *[Henryk IV Probus: Silesian Minnesinger], ibid., 85*

The festival and the attractions of Breslau were promoted on the radio, a relatively new medium in the inter-war period, which quickly became an egalitarian means of communication, thanks to the increasing availability of radio receivers. The Third Reich exploited the great propaganda potential of this small device that was part of household equipment and a link between the highest authorities and ordinary citizens listening to the messages addressed to them. The Breslau radio station (*Reichssender Breslau*) was also involved in organising and running the festival. At the end of June 1937, an hour-long evening programme was broadcast all over the Reich, encouraging everyone to come to the 12th German Choral Society Festival.[8] The broadcast was devoted to the historical and current development of the city, and it was rendered even more attractive by being presented as a kind of reportage, beginning at the tower of the Church of St Mary Magdalene, before listeners were taken to the Town Hall, where they were addressed by Mayor Schönwälder, the royal castle, the university auditorium, to listen to a Mozart quartet, and the cathedral, to hear choral singing. The programme culminated in a visit to the Jahrhunderthalle and an organ concert. The show was designed to promote the city and anticipate the festival, which was to take place in various spaces important to the history and architecture of Breslau. The city was referred to as "Breslau–city in the East!" ("Breslau–Stadt im Osten!") and the "cultural bastion of the German East" ("das Kulturbollwerk des deutschen Ostens"), invoking not only geography, but primarily the geopolitical position of the capital of Silesia. Those are manifestations of the above-mentioned "border discourse", which stimulated the pro-national atmosphere and triggered a sense of threat.

The five days of the music festival

From 28 July to 1 August 1937, from the early morning until late in the evening, Breslau echoed with the voices of tens of thousands of singers. The 12th German Choral Society Festival was attended by choirs, instrumentalists and dance groups from all over the German Reich.

The most important organ concerts, including the grand opening soirée, were held at the Jahrhunderthalle.[9] The other concerts of numerous choirs were held in the open air at the sports stadium in Leerbeuthel, on a sheltered stage, and in the enormous Messehof exhibition hall near the Jahrhunderthalle. Chamber concerts and accompanying events were held in smaller concert halls and public spaces

8 *Breslauer Gemeindeblatt*, 1 July 1937, 213–214. The details of the programme are known only from the press article.

9 Information on the festival programme is based on *Festführer* and *Deutsches Volk*.

throughout the city. Some performances took place in concert halls (Aula Leopoldina, "Capitol" cinema, Konzerthaus at the Zoological Garden) or in buildings which normally served a different function (the assembly room of the Chamber of Commerce, St Mary Magdalene's Church).

The festival programme was focused on traditional folk songs and pieces related to the nation, the military or politics. The Classics were also present, including compositions by Handel and Schubert in the opening concert.

The fifteen short, live tapes from the Breslau festival, performed by different choirs and recorded by Reichssender Breslau, present primarily traditional and folk music, often sung in dialect (e.g. *Jodler aus Vorau*, or *Wir Steirer san recht lust'ge Leut'*).[10]

Another recording is the "Singers' greeting", which was a kind of anthem of the German Choral Society and an interlude during the festival.[11] Although this song does have a text, by an unknown author ("Grüß Gott! Grüß Gott mit hellem Klang! Heil deutschem Wort und Sang!"), this version is instrumental. The jingle for the Breslau festival, both cheerful and solemn, was written by the Silesian composer Ernst August Voelkel (1886–1960), born in Neurode (now Nowa Ruda), Glatz County.[12] It is worth mentioning that Voelkel had worked with Breslau radio station since its founding in 1924. He was one of the "pioneer" composers who developed music forms aimed at the new medium, such as the radio suite and music for radio plays. Most of his works are considered to be lost, so the "Singers' greeting" is very rare.

The festival featured concerts in the city's streets and squares (e.g. Neumarkt, Museumsplatz, Schießwerderplatz, Liebichshöhe; Fig. 4). Even factory workers joined in: on the Friday morning, they were given an hour-long break at work for a singalong in thirty-three factories.

---

10 The recordings from the 12th German Choral Society Festival are held in the Deutsches Rundfunkarchiv in Frankfurt am Main (hereafter DRA), Tonträger DRA Frankfurt, Archivnummer: 4322527.

11 DRA, Tonträger DRA Frankfurt, Archivnummer: 4322072; Deutscher Sängergruß (Pausenzeichen zum Deutschen Sängerbundesfest).

12 Klühs, "Ernst August Voelkel", 5–26. The "Singers' greeting" is not included in the catalogue of his works.

Figure 4: Museumplatz: Konzert des Deutschen Männergesangvereins in Wien *[Museum Square: Concert of the German Men's Choral Society of Vienna], ibid., 71*

Figure 5: Blick auf den Festzug und die Tribünen *[View of the grand parade and stands], ibid., 137*

The festival culminated in a grand parade (Fig. 5), comprising four separate processions moving from different parts of the city toward the city centre and crossing on *Schloßplatz* (now Wolności Square). The columns of people marched through the city for about three hours, singing continuously. The participating choirs were obliged to prepare two or three marching or folk songs, and the choice was left to their leaders. According to the festival guide, the four parades crossed the following streets on their way to Schloßplatz:

1. Kaiser-Wilhelm-Straße (now Powstańców Śląskich Street), via Neueschweidnitzerstraße and Schweidnitzerstraße (now Świdnicka Street).
2. Gräbchener Straße (now Grabiszyńska Street), via Neuegraupenstraße and Graupenstraße (now Sądowa and Krupnicza Streets).
3. Nikolaistadtgraben and Königsplatz (now Podwale Street and Jan Paweł II Square), via Wallstraße (now Włodkowica Street).
4. Trebnitzerplatz (now Powstańców Wielkopolskich Square), via Bismarck- and Heinrichstraße (now Chrobrego and Brodatego Streets), Oderstraße (now Odrzańska Street).

Similar elements (parades and concerts) were also introduced in the German Gymnastics and Sports Festival (*Deutsches Turn–und Sportfest*) held in Breslau the following year. In many ways, that event was similar to the singing festival, and the two events should be seen as closely linked, but due to its sports character it will not be further discussed in this article.

Spectacular parades, meticulously planned, combined visual and audio elements, enriching them with a layer of performance. Besides grand parades at the end of the festival, there were also smaller parades accompanying the opening ceremony and the exchange of banners. The success of these projects was ensured by excellent organisation and discipline, highly praised by the author of one of the festival guides.[13] The grand parades were supposed to reflect, *pars pro toto*, the functioning of the state. As Sallie Marston notes: parades are "very much about monopolizing the space across which a particular imagining of the community can be enacted and projected to a wider audience".[14]

Of course, those actions were both inclusive and exclusive: not all the inhabitants of Nazi Germany could or wanted to feel themselves to be members of the community and take an active part in the events. So on the audio and visual level, we should emphasise the importance of these events as a demonstration of the strength and unity of the National Socialists and their dominance over the rest of

13 *Festführer*, 85.

14 Marston, "Making difference", 383.

society. Special restrictions were imposed on the Jewish population. One of the limitations was a ban on decorating their property with flags and state symbols; they could only be adorned with fresh greenery.[15]

Carolyn Birdsall points also to the phenomenon of "the Festivalisation of the Everyday", through the creation and appropriation of public holidays and sophisticated forms of celebration, which were used to legitimise power and increase support for the party.[16]

The organisation and running of the music festival embodied an exceptionally well considered concept, aimed at dominating the soundscape of the whole city. Some concerts were scheduled in the open air and at venues quite distant from one another. These activities culminated with all the individual voices and smaller groups coming together in a huge choir that sang and marched in the grand parade. Thus the auditory sphere of this event was of key importance for the appropriation of public space in Breslau by the National Socialists.

In his novel, Marcel Beyer recreated a scene of marching and singing people, and captured the atmosphere of the singers' unity:

> The marchers break into a folk song, and the local inhabitants, their cheeks soon hot and flushed, loudly join in the first verse. An entire family sings along, clustered together in a small kitchen window. Clearly visible in their open mouths are tongues, teeth, even threads of saliva. Down below them, noisy expulsions of breath mingle, elbows collide, men jostle one another and break step, their eyelids beaded with sweat in the torchlight. Now they're out of sight. The music fades, the spectators retire into their living rooms. No sound save the agitated twittering of a bird rousted from sleep as it flies across the street. A last, smoldering cigarette butt glows in the night-dark roadway.[17]

At this point, it is important to emphasise the great impact of music and singing, which was appreciated by the authorities of the Nazi Germany. Friedhelm Brusniak, a researcher of the soundscape during the Nazi period, quotes Alfred Rosenberg, who enjoined the members of the German Choral Society in 1934 to "capture the German man, his deep inner strength of mind and will, and all his spiritual attitude". The society, Brusniak continues, was to pursue this tremendous task even when other areas of art lost their impact, because German music and song cherished "the source of eternal renewal that connected millions of people".[18]

Singing in public space was, therefore, all the more desirable as a collective experience that helped strengthen the national community (*Volksgemeinschaft*).

---

15 See point 9 of the guidelines, in *Breslauer Gemeindeblatt*, 21 July 1937, 233.

16 Birdsall, *Nazi Soundscapes*, 65 ff.

17 Beyer, *The Karnau Tapes*, 68.

18 Brusniak, "Der Deutsche Sängerbund", 421.

Soundscape of the future

> The acoustics here are odd. Six microphones are required in front of the speaker's desk alone, four of them for the batteries of loudspeakers aimed at the stadium from all angles. The fifth, which serves to pick up special frequencies, will be adjusted throughout the speech to bring out certain vocal effects. The sixth is hooked up to a small loudspeaker beneath the desk and can be controlled by the orator himself. Additional microphones are installed at a radius of one meter to create a suitably stereophonic effect. Positioning these is an art in itself.[19]

The way Marcel Beyer describes the precise preparations for big party rallies is an adequate metaphor of the overwhelming totalitarian regime, in which everything is rigorously controlled. As Trommler writes, "For the title character Karnau, a technical savant who makes sure that the speeches at Nazi rallies will be intelligible even in large arenas, everything is encoded in sound – a fascinating reflection of the intensity with which the Nazi leadership tried to control the acoustic public sphere".[20]

However, the creation of a soundscape that would serve the transmission and consolidation of specific content to the participants of mass events was not an easy task, as one can observe from the example of the Breslau festival. Difficulties with organising and implementing the propaganda message in the complex and multi-layered urban organism exposed the inadequacy of the existing space in relation to the expectations and programme of the National Socialists.[21]

That was one of the factors apt to seriously affect the so-called general development plans of large German cities, mainly provincial capitals. After initial unsuccessful attempts at small-scale interference in the historical fabric of the city, a decision was made to take more radical steps, involving the demolition of many buildings.

The plans were devised under the law on the urban replanning of German cities (*Gesetz über die Neugestaltung deutscher Städte*), which in 1939 included Breslau.[22] Even before the official inclusion of the city on the programme, the Gauleiter of Silesia, Josef Wagner, hired the renowned architect Werner March, whose designs included the Olympic Stadium in Berlin, to develop urban plans for Breslau.[23]

The issue of Breslau's soundscape acquires special significance when confronted with the parallel process of redesigning the entire city centre. The insufficiency of the

---

19 Beyer, *The Karnau Tapes*, 6.

20 Trommler, "Conducting music", 69.

21 This problem is also described by Joshua Hagen, when he analyses the parades organised in Munich. See Hagen, "Parades", 365.

22 Bundesarchiv Berlin, R113/1029.

23 For more on the plans for Breslau, see Jara, "Wrocławski *waterfront*".

existing urban fabric is clear when we compare the venues of the described festival with the visionary plans for the new spaces in which similar mass events were to take place in the future (Fig. 6). The designs of representative urban spaces in Nazi cities were in fact very theatrical. The assembly square was supposed to be the stage for performances aimed specifically at mass participation. It was a very modern form of participation, whose sources may be traced, as Claire Bishop suggests, to avant-garde theatre practices from the early twentieth century.[24]

Figure 6: Sketch for a re-design of Schloßplatz, Josef Schönwälder, c.1940. © Muzeum Architektury we Wrocławiu, Oddział Archiwum Budowlane Miasta Wrocławia, MAt-1118 syg. 24779

Within the city centre, a leading role was assigned primarily to two areas: the culture forum and the government forum. The culture forum was to be created as a result of the reconstruction of Schloßplatz, which previously served as the assembly square. In contrast, the government forum, the so-called Gauforum, was a completely new urban interior, located near Lessingplatz (now Społeczny

24 Bishop, "Sztuczne piekła", 83–136.

Square), which, along with the office district stretching along the river, introduced a new feature to the north-east of the Old Town.

A special role in mass events was assigned to Schloßplatz, which was the focal point for political propaganda performances graced with the presence of important dignitaries, including Adolf Hitler himself (Fig. 7). On each occasion, the rectangular square was enclosed with stands and occasional decorations, which masked the inconsistent style of the surrounding architecture. The event organisers also managed to find a solution to the small area of the square. The parades marched across it, heading for a different part of the city. In this way, they avoided the crowds and chaos, and the square was continuously "in action". Schloßplatz, no longer the main gathering place, was to be transformed into a culture forum. The concepts for its new spatial development included a new opera house, museum and exhibition pavilion, and the conversion of the royal castle into a hotel.

Figure 7: Der Führer trifft mit seiner Begleitung auf dem Schloßplatz ein. Hinter dem Führer: Reichsminister Dr. Goebbels, Obergruppenführer Sepp Dietrich und Gauleiter Josef Wagner *[The Führer arrives with his escort on Schloßplatz. Behind the Führer: Reichsminister Dr Goebbels, Obergruppenführer Sepp Dietrich and Gauleiter Josef Wagner], reprod. from* Deutsches Volk – Singend Volk..., *132*

The plans for the government district included public buildings for various administrative and party bodies, and some commemorative or symbolic objects. The main feature of the district was a huge assembly square surrounded by representative buildings, the most intriguing of which was the bell tower, usually as part of the planned seat of the Reichsstatthalter, the government plenipotentiary. Almost none of the architectural concepts of the Gaufora were realised except for a partially-built forum in Weimar.[25] The bell towers were to be a dominant new visual and audio feature, like church or town hall towers. The introduction of this element into the architecture of the Third Reich is probably related to fascination with the architecture of fascist Italy and the search for a timeless form in architecture common to both regimes, as well as numerous references to ancient architecture.

The Breslau bell tower would have acted as a counterweight to the towers of the cathedral located on Cathedral Island. In the later concept of December 1939, March moved the bell tower even closer to the river and filled the western frontage of the square with a huge domed building. A forum larger than the Market Square would have enabled large rallies with a sound system to be organised. The planned acoustics of that space was twofold, although in both cases it can be described as overwhelming. Filled with a crowd of enthusiastic Nazi supporters, the square would have made an impact through the mass of people and the sound. On a weekday, by contrast, it would have been a huge, empty and probably silent square, where a stray passer-by would have been able to hear the sound of their own footsteps and the sounds coming from the administrative buildings. The "backstage" of the Gauforum was formed by classicist architecture on two sides, and orderly green areas on the other two sides. Greenery was a common feature of the fora designs, but usually on a smaller scale than was intended in March's project for Breslau. It seems that such a solution was beneficial to the acoustics of the space: instead of reverberating, the sound would be partially absorbed.

Ultimately, neither the government forum nor the culture forum were realised, although in the early 1940s, the designs of individual buildings were still being created and the new government building (now occupied by the Polish provincial administration), the only erected building of the planned administrative district, was under construction.

In the preparation of architectural plans, it was important to recognise the existing space, that is, the "inherited" grid of streets and squares of Wrocław, especially within the context of the new tasks and the vision of an ideal National

25 Wolf, "Weimar".

Socialist city to be realised as a result of a thorough reconstruction. With regard to Nazi architecture, it seems relevant to note Aleksandra Sumorok's view that socialist realist architecture is an "extremely interesting field of observation", on account of "the unique relationship between music and architecture, leading to the creation of a programmed sound universe".[26]

I believe that the experiences of the German Choral Society Festival and similar events played an important role in developing plans for city centre redesign. Although it is not easy to find direct references to this subject in archive records, careful analysis of the sources leads to many interesting observations, from direct auditory elements to more refined forms of creating sound environment and musical culture.

Staccato of the last days

> Throughout the day, the city echoed with a deafening roar and infernal noise caused by the demolition of blocks of flats in the city centre, the volleys of heavy howitzers on Holteihöhe [Kacerskie Hill], the field cemetery and Liebichshöhe, bullets falling from outside, bombs dropped from enemy planes and the whistle of flying machines circling over the city.[27]

The extremely dramatic course of the last days of the German city of Breslau, turned into a fortress, whose authorities delayed the decision to surrender the longest, dramatically changed the soundscape of the city. It seemed inappropriate, to say the least, to remind citizens surrounded by sounds that could bring death at any moment about the pre-war festivals. Yet Gauleiter Hanke and fortress commandant Hans von Ahlfen evoked those events in an appeal to the people: "We remind every native citizen of Breslau of the unforgettable choral society festival and the celebration of sport, when the festively decorated Breslau promised solemnly to the Führer and […] Germans around the world that we would do our utmost to participate in the work of the national and social construction of the great German Reich".[28]

The memory of the music festival must have remained vivid, if it was invoked eight years after the event. Like the sports festival, it was one of the pillars of collective identity for the citizens of the doomed Nazi Breslau. The plans to redevelop the city centre and create new public spaces, which were to form the "sets" for a Thousand-Year Reich, were scrapped. Breslau, dubbed "the cornerstone of the

---

26 Sumorok, "Słyszenie", 211.

27 Peikert, *Kronika*, 39 (excerpt from 20 February 1945).

28 *Schlesische Tageszeitung*, 22 February 1945.

south-eastern part of the German Reich",[29] fell into ruin, to be replaced after the war by the Polish Wrocław.

Bibliography

Beyer, Marcel, *The Karnau Tapes*, tr. John Brownjohn (New York: Harcourt Brace, 1997).

Birdsall, Carolyn, *Nazi Soundscapes. Sound, Technology and Urban Space in Germany, 1933–1945* (Amsterdam: Amsterdam University Press, 2012) (available online: OAPEN library).

Bishop, Claire, *Sztuczne piekła. Sztuka partycypacyjna i polityka widowni*, tr. Jacek Staniszewski (Warsaw: Fundacja Bęc Zmiana, 2015); Eng. orig. *Artificial Hells: Participatory Art and the Politics of Spectatorship* (London: Verso, 2012).

Breslau Statistisches Taschenbuch, Breslau, 1936.

Brusniak, Friedhelm, "Der Deutsche Sängerbund und das *deutsche Lied*", in Helmut Loos and Stefan Keym (eds), *Nationale Musik im 20. Jahrhundert. Kompositorische und soziokulturelle Aspekte der Musikgeschichte zwischen Ost- und Westeuropa* (Leipzig: Gudrun Schröder, 2004), 409–421.

Deutsches Volk – Singend Volk (Berlin: C. A. Weller, 1937).

Fastl, Christian, "Sängerfest", *Oesterreichisches Musiklexikon online*, http://www.musiklexikon.ac.at/ml/musik_S/Saengerfeste.xml, accessed 26 April 2017.

Festführer zum 12. Deutsches Sängerbundesfest Breslau. 28. Juli bis 1. August 1937 (Breslau: Deutscher Sängerbund, 1937).

Hagen, Joshua, "Parades, public space, and propaganda: the Nazi culture parades in Munich", *Geografiska Annaler: Series B, Human Geography*, 2008/4, 349–367.

Jara, Karolina, "Wrocławski *waterfront* w czasach Trzeciej Rzeszy" [Wrocław (Breslau) *waterfront* under the Third Reich], *Roczniki Sztuki Śląskiej*, 25 (2016), 133–152.

Klühs, Gabriele, "Ernst August Voelkel (1886–1960)", in Hans-Jürgen Winterhoff and Helmut Loos (eds), *Fünf schlesiche Komponisten des 20. Jahrhunderts*, Deutsche Musik im Osten, 4 (Bonn: Gudrun Schröder, 1994), 5–26.

Marston, Sallie, "Making difference: conflict over Irish identity in the New York City St. Patrick's Day parade", *Political Geography*, 2002/3, 373–392.

29 "darstellt einen Eckpfeiler im Südostraum des Deutschen Reiches", Breslau Statistisches Taschenbuch, 1936, 32.

Loeser, Martin, and Werbeck, Walter, *Musikfeste im Ostseeraum im späten 19. und frühen 20. Jahrhundert – Rezeption und Kulturtransfer, Intentionen und Inszenierungsformen* (Berlin: Frank & Timme, 2013).

Peikert, Paul, *Kronika dni oblężenia (Wrocław 22 I–6 V 1945 r.)* [Chronicle of a siege (Wrocław 22 January to 6 May 1945)], tr. Karol Jońca, Alfred Konieczny and Stefan Reczek (Wrocław: Zakład Narodowy im. Ossolińskich, 1964); Ger. edn. *"Festung Breslau" in den Berichten eines Pfarrers, 22. Januar bis 6. Mai 1945*, ed. Karol Jonca and Alfred Konieczny (Berlin: Union, 1974).

Sumorok, Aleksandra, "Słyszenie socrealistycznej architektury? Kilka uwag o audiosferze miasta i budynku" [Hearing socialist realist architecture? Remarks on the soundscape of a city and a building], in Jolanta Guzy-Pasiak (ed.), *Muzyka w sztukach wizualnych XIX–XXI wieku* [Music in the visual arts of the nineteenth to twenty-first centuries], *Ikonografia muzyczna. Studia i materiały*, 2 (Warsaw, 2013), 209–226.

Trommler, Frank, "Conducting music, conducting war. Nazi Germany as an acoustic experience", in Nora M. Alter and Lutz Koepnick (eds), *Sound Matters: Essays on the Acoustics of German Culture* (New York: Berghahn, 2004), 65–78.

Wolf, Christiane, "Weimar", in *Gauforen. Zentren der Macht* (Berlin: Bauwesen, 1999), 66–119.

Newspapers

Schlesische Tageszeitung, 27 June 1937, 29 June 1937, 22 February 1945.

Breslauer Gemeindeblatt, 1 July 1937, 21 July 1937.

Andrzej Dębski

Willy Brandt Centre for German and European Studies,
University of Wrocław

From "love in the bright moonlight" to "the corner of dreams": A snapshot of the soundscape of Wrocław in 1945[1]

Abstract: The subject of my analysis is the public soundscape of Wrocław in 1945 as created by cinemas, theatres and entertainment venues. In the chapter, I focus on the activities of these institutions and on the presence of music in the city's soundscape. My analysis also looks at aspects of the post-war co-existence of the Germans, Poles and Russians.

Keywords: Wroclaw, Breslau, soundscape, music, cinema

The subject of my analysis will be the public soundscape of Wrocław in 1945 as created by the cinemas of the time (in the German period, the cinemas functioned as substitutes for other cultural and entertainment institutions, which were closed), as well as theatres and other entertainment venues. I will focus on the activities of these institutions and on the presence of music in the city's soundscape, since the sound of music accompanied the people of the city throughout the year and helped them endure the chores of their daily lives. Since post-war Wrocław was co-inhabited by Germans, Poles and Russians, my analysis will also look at the aspects of the city's soundscape which reflect the post-war co-existence of those three nationalities.

On 1 September 1944, following an order by Joseph Goebbels, all German theatres, *variétés* and other cultural and entertainment institutions closed down, and their staff were mobilised for war. Cinemas, however, were exempt from the order and remained places where both civilians and soldiers could flee from everyday chores and journey to the fictional world of film. Cinemas offered escapism and propaganda, screening musicals, comedies and love stories, as well as the weekly movie chronicle: the *Ufa-Wochenschau*. Some of the closed-down theatres were turned into cinemas, such as the Ufa-Schauspielhaus in Wrocław, launched

1 This article is financed from a project funded by the Polish National Science Centre (Narodowe Centrum Nauki), based on Decision no. DEC-2013/09/D/HS2/00553.

in November 1944 in the building of the former operetta.[2] The staff of the new cinema comprised former employees of the operetta, and some of the actresses were even offered jobs as operators. A proportion of the seats in the auditorium were reserved for workers from the military equipment factory.[3]

The great popularity of cinemas throughout the war is evidenced by statistics. In 1938, Wrocław's cinemas were attended by 7.2 million visitors, while in 1940 the figure rose to 11.3 million, and in the respective years an average resident of Wrocław increased the number of their cinema visits from an average of twelve to eighteen per year.[4] In the following years, the audience continued to grow systematically, as shown by data from the Tauentzien-Theater cinema: in the fiscal year 1938/1939, it was visited by 296,000 viewers, followed by 409,000 the next year, rising to as many as 855,000 in the 1943/1944 season.[5] The cinema operated until 21 January 1945. From 1 June 1944 until its closure, it had a total of 515,000 visitors.[6] Nearly one million people were living in Wrocław at the turn of 1945, an increase from 630,000 in 1939 caused by the relocation of people and industrial plants from western parts of Germany, which also translated into an increase in ticket sales. Statistically speaking, however, each citizen visited the cinema much more frequently during the final years of the war than before the war.

The first Soviet air strike on Wrocław came on 7 October 1944 and resulted in sixty-nine casualties. Until January 1945, however, the city remained relatively peaceful.[7] On 12 January, the Red Army launched a huge offensive from the frontline along the River Vistula, moving towards the Oder. Just a week later, the Russians crossed the former Polish-German border near Namysłów. Following an unconditional order to evacuate the city's civilians (issued on 19 January), 700,000 people left Wrocław within three weeks; 90,000 of them died on their westward journey. On 8 February, the siege operation began, and a week later the ring around the city closed. The act of capitulation was signed on 6 May, but Wrocław continued to suffer from fires for several weeks. As a result of the war,

2 The closure of the two drama theatres in Wrocław (Lobe's in 1935 and Hauptmann's in 1936) triggered a change in the profile of the Schauspielhaus: at the beginning of the 1936/37 season, operetta shows were moved to the Municipal Theatre (Teatr Miejski), whilst "the Schauspielhaus became the only stage in Wrocław for drama" (Majewski, *Teatry dramatyczne Wrocławia*, 52).

3 "Ufa-Schauspielhaus".

4 Dębski, "Kina na Dolnym Śląsku", 223.

5 At that time, the fiscal year ran from 1 June to 31 May.

6 Dębski, "Kina w niemieckim Wrocławiu", 50.

7 Gleiss, "Vor 15 Jahren fielen die ersten Bomben".

ninety per cent of the city's south-western districts were reduced to rubble, with fifty per cent of the Old Town also destroyed. During the siege, 80,000 inhabitants of the city lost their lives.[8]

In light of the above, the Tauentzien-Theater remained open for a long time, only closing after the evacuation, when civilians were leaving the city in "panic and turmoil".[9] It is fair to claim that the majority of the remaining cinemas (nearly forty) also operated until the evacuation order. What were their listings during the first days of 1945? Here is the repertoire of Wrocław's leading cinemas at the time:

1. Capitol: *The Woman of My Dreams* (*Die Frau meiner Träume,* Germany 1944, director: Georg Jacoby);
2. Schauspielhaus and Ufa-Palast: *Sacrifice* (*Opfergang,* Germany 1944, director: Veit Harlan);
3. Tauentzien-Theater: *Back In The Day* (*Seinerzeit zu meiner Zeit,* Germany 1944, director: Boleslaw Barlog);
4. Gloria-Palast: *The Concert* (*Das Konzert,* Germany 1944, director: Paul Verhoeven);
5. Scala: *Happiness in Women* (*Glück bei Frauen,* Germany 1944, director: Peter Brauer).[10]

The films were typically light entertainment productions, with plots focused on the infatuations of the protagonists, with a significant musical context. For example, in *The Concert,* Professor Heink, who tries to beguile young girls, is a famous pianist, and Stefan Hell in *Happiness in Women* is a revue star. The soundtracks played an important role: in the latter film, the viewers watched the couple in love dancing on a theatre stage and singing the title song: "Happiness in Women". In Barlog's film, the story's logline and title were a mother's words to her daughter who wanted to spend a weekend with her friend: "Back in the day when I was young, we also enjoyed spending time together".[11]

---

8 The literature on the siege of Wrocław is extensive, but there is not enough space here to quote it. It is worth noting, however, that on 6 May 2008 and 6 May 2011 conferences on the topic were organised at the University of Wrocław, with the accompanying publications: *Festung Breslau – 1945 – Historia i pamięć* [Festung Breslau 1945: history and memory], and *Festung Breslau – 1945 – Nieznany obraz* [Festung Breslau 1945: an unknown picture].

9 Peikert, *Kronika dni oblężenia*, 15.

10 "Filmtheater".

11 The message in German was enhanced with rhymes: "Seinerzeit zu meiner Zeit war man auch schon gern zu zweit".

The Woman of My Dreams and *Sacrifice* boasted another visual advantage, as they were produced in Agfacolor. According to Tomasz Kłys, the latter of the two films is "an example of an art movie free from any ideological servitude" and dominated by the "atmosphere of the whisper of death", very atypical of German films of the time.[12] The film's screenplay was based on a short story by Rudolf Binding (1912), in which a married man in love with another woman dies of cholera. The film departed from the prose original by depicting the female unmarried protagonist dying and not the married man – a change that resulted from Goebbels's intervention. The propaganda mastermind claimed that the lover should be punished for breaking up an existing marriage. However, the film allegedly contained what Goebbels referred to as the "the erotic of death".[13] Although the shooting was completed in February 1943, the Minister of Propaganda stalled its distribution, and the German premiere in Hamburg had to wait until 8 December 1944. Subsequently, the film was screened in Sweden, where reviewers wrote that it "evoked the memory of exquisite Germany, thanks to its proud and colourful pictures".[14] The colour was undoubtedly a huge asset for the film (Goebbels regarded the visual impression as exceptionally good), but besides the plot and the colour the public may have enjoyed also the carnival ball scene in Düsseldorf, with its accompanying musical track, an echo of the old days when people were able to enjoy such balls. When the film was being shot, such large-scale parties were already prohibited.[15]

The Woman of My Dreams was an even more vivid journey to the good old times. A revue film starring Marika Rökk, an actress often compared to Ginger Rogers,[16] it was also impacted by an intervention from Goebbels, which gave it its final shape. It was completed in September 1943 and was to be screened for the first time around the turn of 1944. However, the propaganda minister did not like the protagonist's allegedly over-frivolous dance steps, so the film was sent back to the atelier in April 1944 and had its premiere in Berlin later, in August. It became an instant hit and earned 8 million German marks within three months, significantly more than the 2.9 million spent on the movie's production.[17] And what did Marika Rökk, the film's revue star, sing about? Here are some of the most popular fragments:

---

12 Kłys, *Od Mabusego do Goebbelsa*, 133.

13 Alt, *"Der Farbfilm marschiert!"*, 286.

14 Beyer, Koshofer and Krüger, *Ufa in Farbe*, 168.

15 Ibid., 174–75.

16 Kłys, *Od Mabusego do Goebbelsa*, 124.

17 Beyer, Koshofer and Krüger, *Ufa in Farbe*, 154.

- "At night, one is not happy to be alone, because love in the bright moonlight is at its most beautiful, you know what I'm talking about, on one hand, the other and all ways besides. One needs a little love. It's love that drives the heartbeat in this world's commotion, on one hand, the other and all ways besides".[18]
- "I'm waiting for you, you're my happiness, and my heart longs only for you! All my dreams end with me at your side; the vision of you is with me always!"[19]
- "Don't look here, don't look there, just look straight ahead, and whatever comes your way – don't worry!"[20]

According to Karsten Witte, this comforting song was a "carpe diem!" message to the audience. In 1944, "world's commotion" (*großes Weltgetriebe*) was the least likely rhyme for the word "love" (Liebe), yet although "the rhyme seemed distant, it was in fact the closest possible".[21] Besides, the sight alone of Marike Rökk dancing and singing must have awakened in the audience the memories of the good old days, when evenings were spent at variety theatres or dance venues.

In his account of the events accompanying the siege, Jakub Tyszkiewicz writes that on 5 May, after three months of fighting, a deadly silence arrived: "The fact that there was no longer any clatter of machine guns or artillery shelling was so extraordinary that it made the Festung's defenders anxious".[22] The silence, without the roar of shelling, must certainly have seemed strange, but the city's soundscape during the siege had not been limited to the sounds of guns alone. There are hints that some cinemas were still working, with shorter or longer hours. For example, here is how Werner Riedel remembered the Capitol in 1954: "The last big success was Marika Rökk with her film *The Woman of My Dreams*. She danced at Christmas and on New Year's Day, and she kept dancing until the Russians began their attack, and then she danced some more when the fleeing peasants were running their cattle herds through the city. During the siege, the defenders often asked for a cinema operator who would play the dancing Marika again. But the frontline

18 "In der Nacht ist der Mensch nicht gern alleine/Denn die Liebe im hellen Mondenscheine/Ist das schönste, sie wissen was ich meine/Einesteils und andrerseits und außerdem/Denn der Mensch braucht ein kleines bißchen Liebe/Grade sie ist im großen Weltgetriebe/Für das Herz wohl das schönste aller Triebe/Einesteils und andrerseits und außerdem".

19 "Ich warte auf dich, du bist das Glück für mich/mein Herz ruft nur nach dir!/Kein Traum geht dahin, den ich nicht bei dir bin/dein Bild ist stets in mir!"

20 "Schau nicht hin, schau nicht her/Schau nur grade aus/Und was dann noch kommt/Mach dir nichts daraus".

21 Witte, *Lachende Erben*, 239.

22 Tyszkiewicz, "Pierwsze dni po 'zmierzchu bogów'", 281.

was closing in on the city and, finally, the foyer was turned into a stable. In March 1945, heavy artillery reduced the building to rubble; yet the theatre survived".[23] A slightly different account of the frontline building was given by Conrad Schumacher (dated 8 May the same year): "The Capitol theatre was burning as we marched by. When we passed it, the entire façade collapsed with a huge roar. Had I and Zunft been 50 metres behind, the rubble would have buried us".[24] It seems, therefore, that the front wall collapsed on 8 May, although the entire building was seriously damaged already in March, as a result of the shelling. The cinema, in turn, functioned until the evacuation, or possibly until March, with sporadic screenings for soldiers who managed to get hold of operators to watch and listen to Marika Rökk one more time.

In his dairy, Max Baselt noted the following under 27 January: "I went to the cinema. Continuous bomb alarms and the first round from the Russian long-range cannons".[25] On 6 February, he wrote: "Illegally, that is, without the necessary pass, I took a number 2 tram to get home to Odertor [now Przedmieście Odrzańskie]. [...] The cinemas are open, the trams are working and they're free of charge – no ticket inspectors!"[26] This is how Peter Bannert remembers the situation in a letter written forty years after the events: "In March, I was able to see *The Woman of My Dreams* at the Gloria-Palast, at the moat".[27] In a note from an investigation dated 7 April, "Senior Corporal Specht left his unit without official leave on 22 March and has not returned since. He did not return from a visit to the cinema on the afternoon of 22 March".[28] Such reports prove that some cinemas were open even in March, towards the end of the month. I do not know of any examples of the cinemas being open during the weeks following the siege (the city experienced particularly heavy bombing at Easter, on 1 and 2 April); however, ad hoc screenings at the time can be confirmed – such as those ordered by the Festung's procurement commander on 12 March: "For the units on leave and where justified by special circumstances, the cinema screening vehicle can be used".[29] The vehicle mentioned was a car with a mobile projector that could be put up and used in any room (before the war, the Silesian seat of the NSDAP had thirty-six such cars; they would

23 As cited in "Das 'Capitol' in Breslau".
24 As cited in Gleiss, *Breslauer Apokalypse 1945*, v:665.
25 As cited in Gleiss, *Breslauer Apokalypse 1945*, vii:593.
26 As cited in Gleiss, *Breslauer Apokalypse 1945*, ii:160.
27 As cited in Gleiss, *Breslauer Apokalypse 1945*, viii:1271.
28 As cited in Gleiss, *Breslauer Apokalypse 1945*, iii:720.
29 As cited in ibid., 383.

travel from one village to another to render "film and propaganda services"[30]). Another order, dated 26 April, provides evidence that films were screened almost until the end of the military operations: "The part of 609th Infantry Division that is in combat shall have at its disposal, by 30 April 1945, 60 seats at movie screenings with sound every day. The exact allocation of seats and the locations and times of the screenings cannot be conveyed in writing due to confidentiality; therefore, the details will be communicated by the national socialist officer in charge of 609th Infantry Division".[31] The vehicle mentioned above was equipped with a 35 mm projector, but narrow tape projectors were also used: it was thanks to the latter that Emil Heinze was able to watch the romantic comedy *Holidays from Myself* (*Ferien vom Ich,* Germany 1934, director: Hans Deppe), on 9 April, at the field hospital.[32]

The orders referred to above pertained not only to films. On 12 March, it was announced that the Festung's entertainment and classical music band composed of five musicians was at the disposal of the injured and those temporarily exempt from frontline action;[33] on 26 April, that apart from the above-mentioned band (conducted by junior officer Wassenberg), the injured could also count on a trio led by senior corporal Philipp. The latter band's name was Colourful Bunker (Bunter Bunker).[34] In addition, Maximiliam Hennig, founder of a string quartet in 1922 and a post-war professor of music in Berlin, was also ordered to play the violin for the injured from time to time.[35]

One of the most unusual musical events to take place in Wrocław during the siege occurred spontaneously towards the end of April: "On a heavily damaged street in the city centre, passers-by, who usually move swiftly through that field of rubble and roofless ruins, suddenly stopped to listen: from the second floor of the ruins of a tenement building, the melancholic sounds of Beethoven's 'Moonlight' Sonata flowed gently down to the street. The artist was playing with perfect mastery, clear skies above him and mayhem around him. He was a soldier who had come across a piano left behind in the destroyed building. He asked his comrades to stand guard at the gate and knew nothing of the street audience that gathered midst the rubble to listen to the concert. He only noticed them only when twenty or thirty people gave him a round of applause and shouted 'bravo'. Surprised and slightly embarrassed by the commotion, he stood in the windowless frame and

30 E. Ro., "Ein Kino fährt ins Dorf".
31 As cited in Gleiss, *Breslauer Apokalypse 1945*, iv:1054.
32 Ibid., 482.
33 Gleiss, *Breslauer Apokalypse 1945*, iii:383.
34 Gleiss, *Breslauer Apokalypse 1945*, iv:1054.
35 Bittner, "Wiedersehen mit Professor Hennig".

bowed, as if the whole setting had been a concert hall – and then he brilliantly played an encore to finish off his performance".[36] Although this account comes from a propaganda frontline gazette, with its specific style (it was also underlined in the article that "difficult wartime experiences have failed to drag us down to a point where we are merely sulking"), the event itself does not seem likely to have been made up by the author. In an exaggerated manner, the author also mentioned lighter music played from gramophones and accordions, which supposedly was to be heard "in all corners of the city".[37]

The Russians began to take up positions in the city on the night of 6 and 7 May; their arrival was accompanied by the "noise of the loudspeakers playing waltzes by Strauss".[38] Germans able to play musical instruments could count on better treatment at the hands of the Russians: "After the capitulation, professor Hennig did not have to go to the prisoner camp at Hundsfeld [now Psie Pole]; as the Russian commander, like many of his compatriots, was fond of music, he ordered that musical talents be spotted among the prisoners in order to later organise a few comforting 'welcome home evenings' for the returning inhabitants of Wrocław. And, by an incomprehensible coincidence, it so happened that Henning was 'engaged' along with a respected opera singer, Kreywanger, to perform at a grand soirée in the hall of the Liebich Theatre, which had survived the siege (although the front building, with its dancing palace and bar, had been destroyed). The soirée was attended by crowds composed not only of Russian forces but also of local civilians, even from distant areas north of the city. [...] The music was appreciated by the Russian units and, from among our soldiers, the artists were first to be freed".[39] The concert at the Liebich Theatre with Hennig and Kreywanger, which most probably took place on 24 June, was also described in a later document of the Antifa anti-fascist freedom movement.[40] Cultural and entertainment activities revived also in other locations in the city. Under 15 July, Horst Gleiss's diary mentions the author taking part in a *variété* performance organised in the yard of the destroyed building of the Scala Cinema. The event included shows by magicians, clowns and dancers supported by a musical trio (although Gleiss was disgusted at the standard of performance, which was supposedly a far cry from the pre-war standards of the Liebich Theatre).[41]

---

36 x., "Unsterbliche Musik".

37 Ibid.

38 As cited in Thum, *Uprooted*, 36.

39 Bittner, "Wiedersehen mit Professor Hennig".

40 Gleiss, *Breslauer Apokalypse 1945*, viii:1576.

41 Gleiss, *Breslauer Apokalypse 1945*, vi:386.

On 9 May, the vanguard of the Polish administrative group arrived in Wrocław, followed the next day by Bolesław Drobner. Very soon, on 14 March, Drobner was appointed Wrocław's mayor by the Temporary Government of the Republic of Poland (under Stalin's patronage). The aim of the Government was to take over power in the city from the Red Army as soon as possible and use the method of a "fait accompli" to "install a Polish administration in German territories east of the Oder-Neisse line prior to the Potsdam Conference".[42] Although the Tehran Conference had reached a preliminary consensus on the issue, the exact course of Poland's western border was detailed at the Potsdam Conference, which took place in July and August 1945. It was only after that conference that the Russian commander, Lyapunov, was able to transfer all administrative power over the city to the Polish mayor. Until then, actual authority was in the hands of the Russians, who permitted the creation of a German administration that dealt with German issues in parallel with the Polish administration.[43] The Germans constituted a clear majority in the city: according to estimates, in mid June there were approximately 2,000 Poles in Wrocław and around 140,000 Germans. At the end of August, the respective figures were 16,000–17,000 Poles and 190,000 Germans.[44] The wave of Polish settlers increased from November, and in February 1946 there were 58,000 Poles in the city among a total of 168,000 inhabitants.[45]

Both Polish and German were spoken on the city's streets and in its shops. In October, the *Pionier* wrote: "At the J. Biegański Colonial Store on Wrocław's Krupnicza Street, to our amazement, when someone addresses the shop assistant, she replies in German: *Ich verstehe Sie nicht.* Just like the other shop assistant. […] Who will ensure at last that a Polish company in the Polish city of Wrocław has Polish staff? It is unimaginable that a Pole should take a dictionary with them when going shopping".[46] Polish as spoken in the city became infused with German loanwords, as in a conversation between two Polish girls overheard and later reproduced in the *Gazeta Dolnośląska*: "*Gutenmor…* oh! Sorry… Good morning Aduś! – pardon me, I've grown so accustomed to these germans that I sometimes forget to speak properly… How did you sleep? Where are you going? – What? To the *wohnungsamt* – You don't say! That's how they had me set up three days ago: they sent me back to the *bürgermeister* and he referred me to the *hausmeister*, and imagine this… Where am I going?… to the *arbeitsamt…* you know! I've got

42 Thum, *Uprooted*, 17.
43 Ibid., 77–79.
44 Kaszuba, *Między propagandą a rzeczywistością*, 42.
45 Suleja, *Historia Wrocławia*, 13.
46 szel, "'Oazy niemczyzny' muszą zniknąć".

a german maid... so intelligent... she gives me german lessons... [...] Besides, just think about it: there's so much business to be done with the germans... I'm looking for a *kutscher.* Yes. It's about transporting those two pianos for Nuśka... [...] Where do we live? Adolf-Hitler-Str. [now Mickiewicza Street]... You? Horst-Wessel [now Wróblewskiego Street]... Oh, great! It's so close! [...] Bye!... *Auf Wiedersehen...* pardon me... bye-bye... sweetie!!!"[47]

The Polish press stigmatised closer contacts with the Germans, which were relatively frequent and not limited to trade. They were also developed at the Scala: "Home-grown crooks and looters are friends with their cronies, Germans. Business, you see, is not politics. It's about buying cheap and selling at a good profit. And people have cash. Go into the Scala cabaret, round the corner from here. See how people eat, dance, have fun and spend cash, man. After hours of trading and selling, this guy's pocket is packed with fat banknotes, he'll go and have a shot and a '*szlofoks*', as he puts it".[48] Although the Scala was visited by both Germans and Poles in October, as evidenced by the quote above, it is hard to say the same of the former Liebich Theatre. In November, it was already operating under the name Teatr Ludowy (People's Theatre), with revues and circus shows. The artists who performed there on 11 November included the dancer Helena Thomas and Galagała, a prestidigitator. The programme featured "interesting shows of acrobatic skills, dance, magic tricks and figure cycling".[49] On 7 December, the 4 Asy (4 Aces), a band continuing the pre-war tradition of Warsaw's Dana Choir, performed on the theatre's main stage. In press reports following that event, the journalist wrote: "Folk songs in an intelligent interpretation, sentimental songs and soldiers' songs provided the audience with the necessary minimum of entertainment, relaxation and satisfaction after their work, something they well deserved after five and a half years of various restraints. [...] The audience also demanded that 'Piosenka o Warszawie' [Song about Warsaw] be sung many times – proof of their simple taste. The 4 Aces sang that song at the end of the show, and they performed with true mastery. One may wonder why a revellers choir of such quality did not perform at the Municipal Theatre instead".[50]

47 "Szanujmy język ojczysty". The word "germans" was intentionally not given a capital G in the newspaper.

48 K., "Spacerkiem po mieście".

49 ha, "Miła impreza w Teatrze Ludowym".

50 Domański, "'4 Asy' w smutnej i wesołej piosence". That passage probably concerned "Kolęda Warszawska" [Warsaw carol], with lyrics by Helena Kołodziejska, first performed on Christmas Day 1944. The 4 Asy included that song on their 1946 album, confiscated by the censors.

The final reflection in the quoted passage was justified, as on 6 November the Municipal Theatre hosted the Jurand Choir (Poland's second most popular choir between the wars after Dana). However, that performance was disappointing: "It was clear on many occasions that a fourth singer was missing, which is why the sound of this 'choir' was incomplete and not harmonised. Moreover, the tremolo representative of the 'sentimental sound', citizen Nader, produced false tones from time to time, which ought to be attributed to his hoarse voice on the day".[51] The Municipal Theatre was a flagship music venue: between 8 and 10 September it hosted Moniuszko's *Halka* and in December Rossini's *Barber of Seville.* The first orchestral concert, conducted by Stefan Syryłło, which opened with a Chopin polonaise, took place at the theatre on 29 June, followed by a concert of Slavic music on 28 July, with the pianist Piotr Łoboz, featuring music by Chopin, Żeleński (a suite of Polish dances), Smetana (*Vltava*), Tchaikovsky (suite from *Swan Lake, Capriccio Italien*), Rachmaninov and Mussorgsky.[52] However, the theatre was also open to events of a lighter character, such as a performance by the Theatre of the 1st Army: "those of us who remember the best revue bands from before the war were able to reminiscence on times spent at the 'Banda' and the 'Cyrulik'".[53] On 2 December, the theatre opened its stage for the "Wesoła Szesnastka [Jolly sixteen] jazz band led by Władysław Rossa, featuring the well-known radio singer Wiktor Ostrowski".[54]

"Jazz" was undeniably in fashion, as is evidenced by announcements made by Wrocław's restaurants from the period. At the Bristol, opened on 4 November, there was a "select jazz band",[55] and at the Bałtyk "a band of top jazz players".[56] It must be noted, however, that the bands did not play "pure jazz", but rather dance numbers "devoid of improvisation".[57] Dance parties were also organised by the Syrena and Arkadia restaurants. Their popularity is shown clearly by a report from the nearby town of Legnica, where the author complained about meagre turnouts at *The Slippers* (*Trzewiczki/Cherevichki*, USSR, 1944, director: Nadezhda Kosheverova, Mikhail Shapiro), a movie version of Tchaikovsky's opera: "There were not many people in the audience. It wasn't hard to find out why. [...] Jazz, that's what you call proper music. Isn't that right? That's why the inhabitants of

51 Cudnowski, "Występ Chóru Juranda".
52 "1. Koncert Popularny w Teatrze Miejskim we Wrocławiu".
53 "Teatr żołnierski". This passage refers to the popular Warsaw-based theatre groups Banda and Cyrulik Warszawski.
54 lj, "Taniec humor i śpiew".
55 Bristol.
56 Bałtyk.
57 Klimsa and Siwek, "Jazz we Wrocławiu", 193.

Legnica decided to go in their droves to one of the trendiest local cafes, where a good jazz band was performing for the first time. [...] The film was excellent, in a nutshell, so it is worth giving up the cafe, the jazz and the passion for swing in order to watch the movie".[58]

Cinemas were also open in Wrocław. The Warszawa (formerly the Palast-Theater) launched on 16 June with a screening of the film *Majdanek – Europe's Cemetery* (*Majdanek – cmentarzysko Europy,* Poland 1944, director: Aleksander Ford). The press reported that of all the pre-war cinemas "only four remained in working order, though with incomplete equipment and in need of renovation".[59] The other three cinemas re-opened as the Polonia (formerly the Odertor-Lichtspiele), on 13 September, the Pionier (Titania-Theater), on 13 October, and the Śląsk (Capitol), on 1 January 1946.[60] Yet even before the Polonia opened, the Wyzwolenie (Liberation) began to screen movies in the gym of the former school in Sępolno, possibly because the area was densely inhabited, as it had not been destroyed to such an extent as the city's southern and western districts.

In a report from 29 August 1945, Ehrenfried Bock wrote that at the Warszawa, taken over by the Poles, Soviet films were screened, and the audience was sizeable only on Sundays. At the Odertor-Lichtspiele, the predecessor of the Polonia under Russian management, some German films found on the premises were also screened, such as *Back In The Day*[61] – an interesting find which may indicate that Soviet wartime movies were not too popular, whereas German escapist movies were more easily digested, even by those of the Russians who did not know any German. However, the explanation may lie elsewhere: Gregor Thum writes about the "unsettling friendship" between the Russians and the Germans and the so-called "Fräulein" phenomenon – German women allegedly entering into relationships with Russian soldiers in order to ensure a higher level of safety and security for themselves.[62] So watching German movies may have been a pleasant pastime for German women visiting the cinemas with their Russian protectors. After all, even though the war was over, the times were still far from normal. At the end of June, a Polish man appealed to the authorities, "on behalf of the peace-loving people of Wrocław", to ban guards and policemen from using guns.[63] Ursula Waage, in turn, draws attention to the terror of the siege and the period after the

58 Domański, "'Trzewiczki' Czajkowskiego w 'Bałtyku'".
59 "Nasze pierwsze kino".
60 Lis, "Kina wrocławskie w latach 1945–1960", 163–64.
61 Dębski, "Kina w niemieckim Wrocławiu", 51.
62 Thum, *Uprooted*, 36–43.
63 W. B., "Czytelnicy piszą".

arrival of the Russians and then the Poles: "When entire buildings are searched by Polish looters, people are making an unearthly noise with pot lids and other kitchen utensils, and shouting the names and numbers of the looted buildings into the darkness of the night. [...] On 16 September, in the Zimpel [now Sępolno] residential area alone, there were 67 daytime and night-time lootings. There are heaps of empty bottles stacked in front of our cellar windows, so every clink of glass sharpens our senses in an expectation of an unwanted visit. We are then relieved to find out that it was only a rat that caused the noise".[64]

The "Fräulein" phenomenon was also present, albeit on a smaller scale, in the relations between Poles and Germans. Interestingly, people would meet at the cinemas, which were probably more attractive for the venue they offered for such dates than for the films they screened. In September, Zygmunt Sztaba wrote: "I'm standing on the bridge. The Oder, King Bolesław the Brave's border river, flows through Wrocław, the city of the Piast dynasty. A few policemen are cycling towards the city. On their way, they pass a group of young German girls who are going in the opposite direction and pulling a small cart packed with tools. The girls wave their hands, laughing and saying '*Hallo!*'. One of the boys shouts: '*Servus Gretel! Nachmittags im Kino*!'. Then they cycle away".[65] One month later, Mieczysław Kolbusz complained: "Fräulein Lisa has a boyfriend from the Industrial Guard. She has fun, sometimes they go to the cinema together (she only needs to take her armband off and whisper, because her 'Stasiek' tells her to do so). They often organise parties at home, and then she invites her two girlfriends".[66]

Regular cinema listings appeared towards the end of November. Earlier screenings can only be reconstructed from random notes in the press or reviews. By the end of 1945, the inhabitants of Wrocław could watch the following movies:

- *Berlin* (USSR 1945, director: Yuri Raisman);
- *In the Name of the Motherland* (*W imię ojczyzny/Vo imya Rodiny,* USSR 1943, director: Vsyevolod Pudovkin, Dmitry Vasiliev);
- *Kutuzow* (USSR 1944, director: Vladimir Petrov);
- *Fire over England* (*Wyspa w płomieniach,* United Kingdom 1937, director: William Howard);
- *Fears* (*Strachy,* Poland 1938, director: Eugeniusz Cękalski, Karol Szołowski);
- *The Medicine Man* (*Znachor,* Poland 1937, director: Michał Waszyński);

---

64 Waage, *Bleib übrig*, 64.

65 Sztaba, "Żołnierz i dziewczyna".

66 Kolbusz, "'Übermensche' bez maski".

- *Ivan the Terrible* (*Iwan Groźny/Iwan Grozny,* USSR 1944, director: Sergey Eisenstein);
- *Far, Far Away* (*Za 7-ma górami,* probable original title: *Kashchei bezsmertny,* USSR 1944, director: Alexander Rou);
- *Antony Ivanovich is Angry* (*Antoni Iwanowicz gniewa się/Anton Ivanovich serditsa,* USSR 1941, director: Nikolay Shengelaya);
- *The Attack* (*Najazd/Nashestvye,* USSR 1945, director: Abram Room);
- *The Folk Avengers* (*Mściciele ludowi/Narodnye mstiteli,* USSR 1943, director: Vasily Belayev);
- *The Sign of the Cross* (*W cieniu krzyża*), USA 1932, director: Cecil B. DeMille;
- *The Great Waltz* (*Wielki walc*), USA 1938, director: Julien Duvivier;
- *Cart Driver 13* (*Dorożkarz Nr. 13,* Poland 1937, director: Marian Czauski).[67]

These films were screened in Wrocław during the last months of the year. If we assume that earlier screenings were also shown in other locations of Lower Silesia, then the list can be extended to include movies such as *The Slippers* and *She Defends Her Motherland* (*Ona broni ojczyzny/Ona zashchishchayet Rodinu,* USSR 1943, director: Friedrich Ermler), *Zoja* (USSR 1944, director: Leo Arnshtam), *The German Defeat at Moscow* (*Pogrom Niemców pod Moskwą/Razgrom nemetskich voysk pod Moskvoy,* USSR 1941, director: Leonid Varlamov, Ilya Kopalin), *Chapayev* (USSR 1934, director: Sergey Vasiliev, Georgy Vasiliev), and *The Rainbow* (*Tęcza/Raduga,* USSR 1944, director: Mark Donski). A large part of the repertoire consisted of Soviet movies, mostly war movies and propaganda productions, but other genres were also included: for example, *Antoni Ivanovich is Angry* was advertised as a "comedy and musical".[68]

The *Pionier* published reviews of selected films, mostly Soviet ones; the only exception was *Fire over England,* a film about the sixteenth-century conflict between England and Spain. The reviewer, Wiesław Domański, liked the movie, as it reminded viewers of the need to put work for the country before personal happiness.[69] Polish films were only mentioned in the paper by means of a jocular epigram: "At the Polonia, they screen *Fears.* Why should I care? At the Warszawa, *The Medicine Man.* That's too much! No reviews here, these movies are too bad to bear!"[70] Despite this criticism, it is highly likely that Polish and American films were included in the repertoires at the end of the year because audiences were

67 This list is based on issues of *Pionier.*
68 "Program kin".
69 Domański, "Sens niezamierzony".
70 Wiesław, "Fraszka".

tired of Soviet productions and needed lighter films to better cope with everyday life. The musical element of films was also important: *The Great Waltz* was a film biography of Johann Strauss II, and the plot of *Fears* revolved around a revue setting. The year’s last film was *Cart Driver No. 13*, with music by Tadeusz Górzyński and lyrics by Jerzy Jurandot.[71] In one of the songs, actors Lena Żelichowska and Mieczysław Cybulski sang the following lines: “There are hours of infinite melancholy. There are bad nights that keep you awake until dawn. That’s when my world is changed, with a ray of light, from the little corner of dreams that I store in my heart”.[72] Although almost everything changed in Wrocław between the beginning and end of 1945, the dreams of Poles watching *Cart Driver No. 13* were probably similar to the dreams of Germans listening to Marika Rökk.

tr. Marta Kałużna-Gołąb

Bibliography

“1. Koncert Popularny w Teatrze Miejskim we Wrocławiu” [The first popular concert at the Municipal Theatre in Wrocław], *Gazeta Dolno-Śląska,* 29 July 1945.

“Arkadja” (announcement), *Pionier,* 6 September 1945.

“Bałtyk” (announcement), *Pionier,* 18 November 1945.

“Bristol” (announcement), *Pionier*, 4 November 1945.

“Das ‘Capitol’ in Breslau”, *Der Schlesier* 8 (1954).

“Filmtheater”, *Schlesische Zeitung*, 29 December 1944.

“Nasze pierwsze kino” [Our first cinema], *Nasz Wrocław,* 17 June 1945.

“Program kin” [Film listings], *Pionier,* 30 November 1945.

“Repertuar kin” [Film listings], *Pionier,* 30 December 1945.

“Syrena” (announcement), *Gazeta Dolno-Śląska,* 15 August 1945.

“Szanujmy język ojczysty” [Let’s respect our native tongue], *Gazeta Dolno-Śląska,* 26 August 1945.

“Teatr żołnierski” [The military theatre], *Gazeta Dolno-Śląska,* 15 July 1945.

“Ufa-Schauspielhaus”, *Schlesische Zeitung*, 7 November 1944.

Alt, Dirk, *“Der Farbfilm marschiert!”. Frühe Farbfilmverfahren und NS-Propaganda 1933–1945* (Munich: Belleville, 2013).

71 “Repertuar kin”.

72 “Są godziny melancholii bez dna/Są złe noce nieprzespane do dnia/Wtedy mi odmienia świat/Opromienia świat/Ten maleńki kącik marzeń w sercu mym”.

Beyer, Friedemann, Koshofer, Gert, and Krüger, Michael, *Ufa in Farbe. Technik, Politik und Starkult zwischen 1936 und 1945* (Munich: Collection Rolf Heyne, 2010).

Bittner, Erwin, "Wiedersehen mit Professor Hennig", *Der Schlesier*, 20 (1958).

Cudnowski, Henryk, "Występ Chóru Juranda" [A performance by the Jurand Choir], *Pionier,* 8 November 1945.

Dębski, Andrzej, "Kina na Dolnym Śląsku dawniej i dziś" [Cinema in Lower Silesia past and present], *Dolny Śląsk,* 19 (2016), 220–28.

– "Kina w niemieckim Wrocławiu (1906–1945)" [Cinema in German Wrocław], in Andrzej Dębski and Marek Zybura (eds), *Wrocław będzie miastem filmowym… Z dziejów kina w stolicy Dolnego Śląska* [Wrocław will be a cinema city… From the history of cinema in the capital of Lower Silesia] (Wrocław: Gajt, 2008), 29–54.

Domański, Wiesław, "'4 Asy' w smutnej i wesołej piosence" [The 4 Aces in a sad and gay song], *Pionier,* 12 December 1945.

– "'Trzewiczki' Czajkowskiego w 'Bałtyku'" [Tchaikovsky's *The Slippers* at the Bałtyk], *Pionier,* 23 September 1945.

– "Sens niezamierzony" [Unintented meaning], *Pionier,* 24 October 1945.

E. Ro, "Ein Kino fährt ins Dorf", *Breslauer Neueste Nachrichten*, 27 March 1938.

Gleiss, Horst, "Vor 15 Jahren fielen die ersten Bomben. Zum ersten sowjetischen Luftangriff auf Breslau am 7. Oktober 1944", *Der Schlesier* 40 (1959).

– *Breslauer Apokalypse 1945*, ii (Rosenheim: Natura et Patria, 1986).

– *Breslauer Apokalypse 1945*, iii (Rosenheim: Natura et Patria, 1987).

– *Breslauer Apokalypse 1945*, iv (Rosenheim: Natura et Patria, 1987).

– *Breslauer Apokalypse 1945*, v (Rosenheim: Natura et Patria, 1988).

– *Breslauer Apokalypse 1945*, vi (Rosenheim: Natura et Patria, 1990).

– *Breslauer Apokalypse 1945*, vii (Rosenheim: Natura et Patria, 1993).

– *Breslauer Apokalypse 1945*, viii (Rosenheim: Natura et Patria, 1995).

Głowiński, Tomasz (ed.), *Festung Breslau – 1945 – Historia i pamięć* [Festung Breslau 1945: history and memory] (Wrocław: Gajt, 2009).

– *Festung Breslau – 1945 – Nieznany obraz* [Festung Breslau 1945: an unknown picture] (Wrocław: Gajt, 2013).

Ha, "Miła impreza w Teatrze Ludowym" [A pleasant event at the Popular Theatre], *Pionier,* 13 November 1945.

K., "Spacerkiem po mieście" [A stroll around town], *Pionier,* 13 October 1945.

Kaszuba, Elżbieta, *Między propagandą a rzeczywistością. Polska ludność Wrocławia w latach 1945–1947* [Between propaganda and reality: the People's Poland of Wrocław 1945–1947] (Warsaw: PWN, 1997).

Klimsa, Bogusław, and Siwek, Wojciech, “Jazz we Wrocławiu. O początkach festiwalu Jazz nad Odrą” [Jazz in Wrocław: on the beginnings of the Jazz on the Oder festival], *Dolny Śląsk*, 19 (2016), 192–204.

Kłys, Tomasz, *Od Mabusego do Goebbelsa. Weimarskie filmy Fritza Langa i kino niemieckie do roku 1945* [From Mabuse to Goebbels: Fritz Lang's Weimar films and German cinema up to 1945] (Łódź: Wydawnictwo Uniwersytetu Łódzkiego, 2013).

Kolbusz, Mieczysław, “‘Übermensche’ bez maski” [“Übermensche” without masks], *Pionier*, 27 October 1945.

Lis, Piotr, “Kina wrocławskie w latach 1945–1960” [Wrocław cinemas 1945–1960], in Sylwia Kucharska and Karol Jachymek (eds), *Wrocław i film* [Wrocław and film] (Warsaw: Wydawnictwo SWPS, 2012), 158–74.

lj, “Taniec, humor i śpiew” [Dance, humour and song], *Pionier*, 1 December 1945.

Majewski, Tomasz, *Teatry dramatyczne Wrocławia w okresie rządów narodowosocjalistycznych 1933–1944* [Playhouses in Wrocław during the national socialist era 1933–1944] (Wrocław: Atut, 2003).

Peikert, Paul, *Kronika dni oblężenia (Wrocław 22 I - 6 V 1945)* [Chronicle of the siege (Wrocław 22 January to 6 May 1945)] (Wrocław: Zakład Narodowy im. Ossolińskich, 1964).

Suleja, Włodzimierz, *Historia Wrocławia* [History of Wrocław], iii: *W Polsce Ludowej, PRL i III Rzeczypospolitej* [In the People's Poland, the People's Republic of Poland and the Third Republic] (Wrocław: Wydawnictwo Dolnośląskie, 2001).

szel, “‘Oazy niemczyzny’ muszą zniknąć” [The German ‘oases’ have to go], *Pionier*, 28 October 1945.

Sztaba, Zygmunt, “Żołnierz i dziewczyna” [The soldier and the girl], *Pionier*, 26 September 1945.

Thum, Gregor, *Uprooted. How Breslau Became Wrocław during the Century of Expulsions* (Wrocław: Via Nova, 2011).

Tyszkiewicz, Jakub, “Pierwsze dni po ‘zmierzchu bogów’” [The first days after the “twilight of the gods”], in Tomasz Głowiński (ed.), *Festung Breslau - 1945 - Historia i pamięć* [Festung Breslau 1945: history and memory] (Wrocław: Gajt, 2009), 281–286.

W. B., “Czytelnicy piszą” [Readers' letters], *Nasz Wrocław*, 24 June 1945.

Waage, Ursula, *Bleib übrig. Aus den Tagebuchaufzeichnungen in der Festung Breslau und der Nachkriegszeit von Januar 1945 bis April 1947 (Zeitzeugenbericht)* (Halle: Cornelius, 2004).

Wiesław, "Fraszka. Zamiast recenzji" [A trifle: in lieu of a review], *Pionier,* 8 November 1945.

Witte, Karsten, *Lachende Erben, Toller Tag. Filmkomödie im Dritten Reich* (Berlin: Vorwerk 8, 1995).

x, "Unsterbliche Musik", *Schlesische Tageszeitung*, 1 May 1945.

Renata Tańczuk

Institute of Cultural Studies, University of Wrocław

The 1945 soundscape of Wrocław in the accounts of its post-war inhabitants

Abstract: The story of the changes in Wrocław and its soundscape in 1945 was written down by a number of earwitnesses. The analysis of sound representations of the city and its distinctive sounds contained in autobiographical accounts, memoirs and diaries, and also in film and radio productions, allows us not only to observe how the identity of the city is constructed, but also to track the process of identification with a place through the ephemeral, embodied auditory experience of that place.

Keywords: transformation of the urban soundscape, autobiographical accounts, auditory topoi, narrative strategy

The 1945 soundscape of Wrocław in the accounts of its post-war inhabitants

The year 1945 was definitely the most disastrous in the history of Wrocław. The death toll reached tens of thousands of civilians and defenders, and the urban fabric was almost totally destroyed.[1] The Festung Breslau memoirs, documents and photographs compiled by Horst Gleiss in ten volumes were not without reason entitled *Breslauer Apokalypse 1945*.[2] The period of time from January 1945 till the surrender on 6 May 1945 was described by witnesses to the Battle of Breslau, regardless of their nationality, with one word: *hell*.[3] The siege and the tactics of

---

1 It is estimated that the defence of the fortress, which lasted for over eighty days, cost the lives of about six thousand soldiers and probably up to eighty thousand civilians. The city was almost totally destroyed and covered with eight million cubic metres of rubble and ash. 90% of the buildings in the southern and western districts and 60% of the old town, including the islands on the Odra River, were wiped out. None of the tramlines or the ten railway stations was usable. Almost all the facilities indispensable to the proper functioning of the city were devastated: street lighting (100%), hydroelectric plants (25%), gasworks (60–70%) and power plant installations (60%). 60% of Wrocław's factories were torn down. See Majewski, *Wrocław-godzina*, 203–204.

2 Gleiss, *Breslauer Apokalypse 1945*.

3 See the accounts contained in Majewski, *Wrocław-godzina*, 20, 137, 154; Kosmulska, *Niewolnicy Breslau*, 146, 245, 256.

the German defence command turned the city into a pile of rubble, a space that was to gain a new national attribution and new inhabitants.

The story of the transformation of Wroclaw and its soundscape in 1945 was reported by a number of earwitnesses in three languages: German, Polish and Russian. It starts with testimonies given by the indigenous inhabitants, who portrayed the process of the terrible destruction of the city and its community.[4] A particularly impressive testimony was given by one anonymous German witness, who described the tragic bombing of the city on Easter Sunday 1945 most acoustically and vividly:

> The air is filled with continual roaring, howling, bursting, there's the thunder of aircraft and the whistling of shells. On the top of that, there's shelling from the south, east and west. Death has a rich harvest and celebrates with orgies of destruction. The bunker seems to rise up and fall back down with a crash. The building shakes like a mortally wounded animal. We are stunned by the air pressure and put our hands over our ears so we don't have to listen to this terrible, hellish symphony any more, and ram our fists into our mouths so we do not scream out of mad anxiety. Some fall to their knees and pray or mumble some incomprehensible noise to themselves. Some seem powerless out of horror. And everyone's face is ashen and looks very old. Hundreds of aircraft seem to be over our block alone, turning in circles continuously, and in the distance we can hear the endless, endless roaring, rising and falling, the melody of death.[5]

Polish refugees and forced labourers living in Breslau during the last year of the war also described the tragedy of the city and the excruciating experience of the final fight.[6] It is worth quoting here two excerpts from the memoirs of Irena Siwicka, who worked at a labour camp hospital for foreigners during the siege and gave an account of the phonic dimension of the nightmare of the bombardments:

> The camp commandant set up a music band. It made an incredible impression when, amidst the most massive bombing and the horror of fetching casualties, he would walk the stairs of all the floors and lead the orchestra playing lively tunes. Before Easter, women would wail "Gorzkie żale"[7] wistfully and excruciatingly at one end of the corridor, and the orchestra would play noisily and cheerfully at the other. The laughing commandant

4 It would be interesting to analyse the vocabulary used by the inhabitants of Breslau to describe the destruction of the city. Accounts of the bombing on the critical day of Easter Monday 1945, for example, included expressions like "firestorm" (*Feuersturm*) and "sea of flames" (*Flammenmeer*). See Hargreaves, *Hitler's Final*, 175.

5 Hargreaves, *Hitler's Final*, 172.

6 Not only did they describe the destruction of Breslau, but they also repeated how beautiful and metropolitan the city was. See Damczykowa, "Znak 'P' nosiłam", 81, 88; Klatko, "Kupił nas Niemiec", 153; Majewski, *Wrocław-godzina*, 217.

7 "Bitter Laments", an ancient Polish hymn sung weekly during Lent.

> would watch them while stamping his feet and clapping the rhythm. It looked as if we were in a sort of an absurd madhouse.
>
> [...] For some time, at exactly 8 pm, wonderful lanterns were hung over the barracks. It was as bright as day. Somewhere nearby, bombs were falling. Aircraft were in combat overhead. Searchlights crossed one another, their colourful balls of light sweeping the sky. The sight was as menacing as it was beautiful. When the rumble subsided and only in the distance were new fires and wreaths of smoke visible, nightingales would start trilling in nearby bushes. The nightingales' singing, the charm of the spring night, the moon disinterestedly watching the human folly, and the horror of the moments just lived through – they were so tightly intertwined that it is hard to understand how one can receive such dissimilar sensations simultaneously. The day on which the massive full-scale attack on Wrocław was launched was horrible. It seemed like all the cannons in the world were booming, that the aircraft of the whole world were dropping bombs, that every centimetre of the air was filled with bullets and shrapnel. The camp was being shot through. Shells were flying above the barracks. The thin wooden walls were creaking and shaking. Smoke was stinging our eyes. Nothing could be done, because shrapnel was whistling through the operating room.[8]

There are also memories of Poles who recounted the very end of the war. Jerzy Kuzior, who in 1943 was forced to work for Linke-Hofmann Werke (a railway car manufacturer), remembers:

> Finally, on Saturday 5 May, a strange silence lying over the city indicated that capitulation talks had started. On 6 May, after three months of destructive fighting, fortress Wrocław capitulated, and on the night of 9 May – we were still living on the allotments – a powerful fusillade of shots announced the end of the war.[9]

And Wanda Urbańska, who from 1943 worked at Regina Weinstube (a wine bar and restaurant) on Dorotheengasse (ul. Św. Doroty), wrote:

> At last the longed-for day came – 6 May. The shots died away, a hush descended. This time it really was the end. In the evening, they told us to go wherever we wanted to. I was walking all night in the direction of the city centre, through rubble and ashes. On the way, I saw columns of German troops. They were walking without belts, with their heads down, lacerated. I went back to the cellar where I had left my personal belongings. There were a lot of German families there. Now they were greeting me with smiles, yet some didn't believe that they had been defeated. They were in despair, they cried, and I was happy. In the morning, a few Russian soldiers entered our cellar. I talked to them as liberators. But soon I was bitterly disappointed, as the liberators pulled off my ring and my watch and raped a German woman right in front of me. I couldn't believe my eyes!

---

8 Siwicka, "Szpital", 260, 268–269.

9 Kuzior, "Moje pierwsze", 77.

> I was horrified again. I covered my dark curly hair with a headscarf. Now my only wish was to get out of the city and return to my hometown, Wieluń.[10]

In some respects, as far as experiencing the city is concerned, Breslauers and Wrocławians were in a similar situation. The former were faced with the unfamiliar experience of living in a city that had been bombed, demolished, burnt down and deserted, roamed through by masses of refugees, a city in the process of radical transformation, which suddenly became dangerous and "lifeless", and they themselves were about to be uprooted.[11] The latter started coming to Breslau on 9 May 1945,[12] and they found the city utterly foreign. The newcomers comprised a heterogeneous group. They came mostly from villages and small towns, and for many of them living in a big city, and also the life of a big city, was an unfamiliar experience. After the war, Wroclaw lost its metropolitan character for many years, and it was derisively called "a big village with trams".[13] The settling of people from rural communities, who insisted on continuing their rustic lifestyle, resulted in the "strong de-urbanisation of [Wrocław's] character", its rusticalisation.[14]

There is copious literature on the transformation of Wroclaw in and after 1945, yet little or no attention has been paid to the auditory experience of its inhabitants, the changes in the city's soundscape and its reception at the moment of the disruption and later. Perhaps we are not yet convinced of the importance of recording and analysing such changes in their relation with the supposedly more noteworthy transformation of the visual landscape of the city, and especially of its industrial, social and economic character. The analysis of representations of Wrocław and its distinctive sounds contained in autobiographical accounts, memoirs and diaries, and also in film and radio productions, allows us not only to observe how the

---

10 Urbańska, "Ukradziono mi", 134–135.

11 First, gauleiter Karl Hanke forced them to leave the city in January, which ended in tragedy: 18,000 people froze "during the Kanth [Kąty Wrocławskie] death march" (Hargreaves, *Hitler's Final*, 214). Then, after the war and the decision to repatriate Breslauers, they were being relocated from October 1945 till 17 December 1946, when the last train with the expellees departed Świebodzki Railway Station (ibid., 242).

12 On that day the first lorry bringing 'pioneers' from Cracow arrived in Wrocław.

13 Zaborowski, "Zamiast Białogard", 178.

14 Thum, *Obce miasto*, 139. Thum devoted an entire subchapter (pp. 139–146) to the problem of rusticalisation; "with the displacement of the German inhabitants of Wrocław, the cultural core that maintained the urban lifestyle was disappearing. Thus, quite suddenly, the social basis for urbanisation – the marker of the patterns and norms of the metropolitan lifestyle for immigrants from the countryside – was missing" (ibid., 140).

identity of the city is constructed, but also to track the process of identification with a place through the medium of sounds, which are ephemeral and yet most important for the embodied experience of a place.

This paper represents an attempt to present the auditory reception and experience of Wroclaw in 1945 found in the accounts of its first Polish inhabitants. Having in mind the significance of sound and the sonic dimension of the experience of a place for local identity, it is worth highlighting what sounds were registered by those "pioneers", what their reception was like, and what meanings, and perhaps values, were attached to them. The tools that can be adopted for the analysis of these phenomena and processes were developed by a research group led by Karin Bijsterveld which studied the ways in which the sounds of a city are "staged" and dramatised in texts, radio broadcasts and films. Those tools were presented in the Introduction to *Soundscapes of the Urban Past* and elaborated in the first chapter of that book by Karin Bijsterveld, Annelies Jacobs, Jasper Aalbers and Andreas Ficker. The researchers identified four ideal types of auditory topoi:[15] *comforting sound* (bringing a feeling of harmony and security), *intrusive sound* (disturbing, threatening), *sinister sound* (which can be ominous and haunting) and *sensational sound* (the positive opposite of intrusive sounds; for example, the familiar noise of traffic or crowds). Their research tools were complemented by the categories of *keynote sound, sound mark* and *sonic icon,*[16] and by the *narrative strategies* employed by the authors of accounts most frequently to imply their sound experience and the impression of the urban soundscape:[17] "arriving in the city, following the rhythm of the urban day and juxtaposing soundscapes in terms of either space (different neighbourhoods within a city) or time (past and contemporary soundscapes)", and "adaptation to sound".[18]

Arriving in the city

At first glance, analysing Wrocław memoirs and diaries is quite disappointing. There are only a few urban soundscape descriptions that can be found in the recollections of pioneers compiled by Włodzimierz Suleja and Mieczysław Markowski, in a collection of stories gathered by Bohdan Jałowiecki, in *Pamiętniki osadników Ziem Odzyskanych* [Memoirs of the pioneers of the regained territories], in Joanna Konopińska's *Tamten wrocławski rok 1945–1946. Dziennik.* [That Wroclaw year

---

15 Bijsterveld, Introduction, 13, 18–21.

16 Ibid., 15–16.

17 Ibid., 15; Aalbers et al., "Shifting", 35.

18 Aalbers et al., "Shifting", 36.

of 1945–1946. A journal] and in Andrzej Jochelson's *Kronika: Semipałatyńsk-Wrocław.* [Semipalatinsk–Wrocław chronicle].[19] The authors seldom register sounds; they are preoccupied with reporting the process of putting down roots, with meetings and activities, with the appearance of the city and the process of its reconstruction. Sounds are not absent altogether, yet the fact that they are so rarely registered prompts one to ponder the circumstances conducive to noticing them and study the way their hearing was related.

It seems that we pay special attention to sounds when we explore a new space, alert to danger, and also when the sonic ambient changes considerably, the soundscape of a place fails to meet our expectations or the acoustic phenomena are tiring and unacceptable. Such situations are among the most important factors that trigger the mode of attentive listening. When we live in a certain space for a longer time, when nothing special happens in our surroundings and we feel safe, the soundscape is registered in our journals and memoirs rather obliquely – we can only presume that some sounds are present, because we know that the described people and things produce them. Then we most often notice those sound phenomena which are unexpected and startling, which interfere with the habitual soundscape or are connected with festivals and celebrations in the urban space. We also perceive the sonic events that we take part in, listen to or find important for some reason (e.g. symbolically, aesthetically).

For obvious reasons, the first of the above-mentioned narrative strategies, *arriving in the city*, the motif of meeting Wrocław, was most often used in the analysed memoirs and diaries. People arriving in the city are listening to it. Through

---

19 The following publications were analysed: Jałowiecki (ed.), *Związani z miastem*; Dulczewski and Kwilecki (eds), *Pamiętniki*; Markowski (ed.), *Trudne dni*; Jochelson, *Kronika*; Konopińska, *Tamten wrocławski rok 1945–1946*; Suleja (ed.), *Wspomnienia.* Some of these publications are journals which were edited and corrected by their authors before publication (e.g. Konopińska), some were published unchanged (Jochelson), others were written to order (e.g. the 1966 writing competition "What is Wrocław for You?" – Jałowiecki's collection). They date from different years and are of various form and content (journals, memoirs, competition stories), and all these factors determined their narrative strategies. Regardless of the reason for and the time of their writing, the recalled components of the urban soundscape reveal some similarities in all the analysed texts. When analysing those of the sources that were written down a long time after the evoked events, it would be worth taking into account the function of sound memory. These issues, however important, have not been the focus of my studies, which, partly because of the dearth of materials, were limited to identifying auditory topoi and narrative strategies. For more on "sound memory", see Birdsall, "Earwitnessing"; Birdsall, "Sound memory"; Bijsterveld, "Beyond".

its sounds, they are trying to figure out what it is like and if it is inhabitable. Wrocław greets them with the sounds of gunfire or poignant silence:

> The city was quiet and calm. No noise of voices was coming from anywhere. From time to time, the silence was broken by a burst of shots from an unknown direction. But it was a sign that people were there, that they were alive. That the city was not dead altogether.[20]

> We get out of the car. We send out pickets in three directions to explore the passages, while we look around the empty streets. The wonderful spring weather is in stark contrast to the sight all around. Destruction, dead silence and desolation everywhere. Here and there only groups of plunderers bustle about.[21]

The silence registered by one of the authors can be identified as intrusive or sinister, but the sounds of shots are ambivalent. On one hand, they are foreboding, portending danger; on the other, paradoxically, they herald the hoped-for presence of other people, so they have a comforting potential.

One of the first to arrive in Wrocław was Kazimierz Kuligowski,[22] who noted intrusive sounds:

> Wrocław is on fire, its streets lie under the rubble of burning houses. [...] One can hear the thwack of burning ammunition exploding. We lose our way in the burning streets, weaving in and out; it seems there is no way through.[23]

> The first and the following nights will always remain in our memory, because they were deceptively reminiscent of what we lived through in burning Warsaw. Nights brightly lit with flames, streets enveloped in smoke and falling ash, explosions of mines, shells and ammunition, hollow cracks of rifles. In a word: a full concert of war.[24]

Andrzej Jochelson,[25] one of the most prominent pioneers, arrived in the city at midnight on 28 May 1945 "to complete the great, apparently impossible, task of Polonising Wrocław".[26] According to his account, till the end of July, fires were continuously breaking out, and one could hear the sounds of gunfire and of looters. This is what he wrote when he first saw the city:

---

20 Pilak, "Kocham w nim", 84.

21 Kulczyński, "Grupa", 111. Stanisław Kulczyński, 1895–1975, a botanist, professor and vice-chancellor of the universities of Lviv and Wrocław.

22 Kazimierz Kuligowski became deputy mayor of Wrocław in 1945.

23 Kuligowski, "W ruinach", 142.

24 Ibid., 144.

25 Andrzej Jochelson (1911–1997), a lawyer, writer, translator and member of Wrocław authorities.

26 Jochelson, *Kronika*, 188.

> Already on the outskirts of the city, we caught sight of a house on fire: it was burning like a warning to us that life would not be easy there. Then we started to skirt around the barricades. The bus's huge body could hardly get through the narrow gaps. So naturally we all got off, to relieve our good old stagecoach. After crossing two barricades, we realised that none of the travellers knew which way to go. The Cracow Office of Wrocław Administration had appointed the leaders of the expedition so cleverly that nobody knew the city. So the driver decided that we would spend the night in the street [...]. The night was very pleasant, warm and full of charm; nightingales were singing. But that did not calm the mood among the members of the team, who, except for me, set about heartily looting: stealing the things left by the Germans. So everybody was looting, from university professors to Mikołajczyk, the cook, who was the last to get to it... Everything was taken: books, underwear, crockery, tooth powder, aftershave, even dummies.
>
> It was an annoying sight for me, to say nothing of the looting noise that kept me awake.[27]

In his narrative, gunfire and explosions were everyday sounds that accompanied the old and new inhabitants of Wrocław. The city is described with the use of intrusive sounds, but also sounds that can be recognised as the sound marks of cities in a state of war or anomie.

Jochelson remarks on 3 June 1945:

> Wrocław is continually on fire, and gunfire can be heard all the time. Everyday there are two, three or even four fires. The skeletons of buildings set on fire by highly inflammable liquids burn down to ashes within hours. At least one plume of smoke can always be seen on the horizon. Apparently, the fires are lit by the Germans, the Nazis. Gunfire is also often heard, either from one direction or an exchange of fire.[28]

Under the date of 10 June, there is another mention of "the traditional sounds of Wrocław": "gunfire and exploding mines, interspersed with singing nightingales, chirping swallows and tweeting sparrows".[29] The dissonant concurrence of the sounds of the city is worth noting here. The mingling of the discordant sounds and soundscapes has tremendous potential for highlighting the drama of the described situation and the horror of living in the city. It can be recognised as the third narrative strategy listed by the authors of *Shifting Sounds*: juxtaposing soundscapes. This strategy also appears in accounts from the period of *Festung Breslau*, including Irena Siwicka's account cited at the beginning of this article and Hugo Hartung's diary entry of 19 February 1945:[30]

---

27 Ibid., 193–195.

28 Ibid., 201.

29 Ibid., 209.

30 Hugo Hartung (1902–1972), a German dramatist and writer, in 1945 conscripted into a Breslau Luftwaffe ground unit.

> with the sunrise, the Russian artillery wakes up too and starts shelling our positions. A grenade strikes less than fifty metres from us, showering the trench with shrapnel. The rails are clanking under the drumfire of machine guns. The most unpleasant thing is the hail of the railway embankment gravel drumming against our helmets after every burst of fire. But the highest note in that hellish concert is maintained by the first spring lark soaring through the skies and singing blithely.[31]

Remarkably evocative descriptions of discovering Wrocław can be found in the 1945–1946 journal of a twenty-year-old writer, Joanna Konopińska.[32] One of the main themes of her journal is the experience of foreignness, the emotions associated with taking over places and touching somebody else's belongings. She also took note of sounds entirely different from those quoted above. On the third day after her arrival in Wrocław, 12 June 1945, "at the table in somebody else's house in a foreign city",[33] she describes the way to her new home in Heinzelmännchenweg (now Henryka Rodakowskiego St.). Her account contains a harrowing portrayal of a devastated city: destroyed houses, broken windows, bomb craters, strewn paper and photographs, a fallen tram, an upturned piano, military cars. Opposite the Jahrhunderthalle (Centennial Hall), she notices the zoological garden:

> Behind the enclosure, two zebras could be seen walking. Somebody told my father that the elephants had been eaten during the siege of Festung Breslau, but I don't know if that's true. Some unseen animals were roaring, perhaps from hunger or thirst. In such a wilderness, that made an incredible impression.[34]

A few pages later, she remembers the first evening at home:

> After supper, which consisted of bread roasted in oil, we were sitting on the balcony. It grew dark. There was a light breeze, which was swaying the tree branches, and the scent of roses was coming from the garden. It was very quiet and calm. It seemed to us that we were all alone in the dead city.[35]

With time, trains loaded with repatriates arrive in Wrocław. They fill the city with new sounds, typical of the countryside. On 5 January 1946, Konopińska recorded the bustle and noise of the unloading of trains:

> Along Psie Pole railway station, there was a long line of freight carriages: the repatriates' property was being unloaded. Somebody was calling somebody else, a mother was looking

---

31 Hartung, *Śląsk*, 73–74.

32 Joanna Konopińska (1925–1996) also wrote *We Wrocławiu jest mój dom. Dziennik z lat 1946–1948* [My home is in Wrocław. Journal of 1946–1948].

33 Konopińska, *Tamten wrocławski*, 36.

34 Ibid., 39.

35 Ibid., 42.

> for her child, driven cattle were mooing, hens were clucking in their cages, bundles of bedding, agricultural machinery and domestic appliances were scattered around everywhere. Commotion, confusion and exasperation could be observed at every turn. On many faces, one could see an expression of doubt and fear of the unfamiliar, the great unknown, what life would bring to them. And the snow was falling calmly, covering the furrowed and trampled earth with a white layer.[36]

The author took down a conversation with a woman, one of those who had come from Stanisławów (now Ivano-Frankivsk, in western Ukraine) and asked the oft-repeated question: "Can one live here? Tell me if one can live here".[37]

Adaptation

During the first months after the war, Wrocław was a dangerous city, in which people were killed and raped. A sexual assault was heard by Jochelson:

> I idled away the afternoon in the office. In the evening, in Szczytnicka Street, I saw a raped girl going back home with her father. She was crying loudly. Under her skirt, a scrap of her knickers was hanging down. There were feathers in her hair. She was leaning on her father. The tragedy of that couple, as a whole, made a powerful impression. She must have been raped by Soviet soldiers. Such incidents are quite common. Also at night, at the beginning of this week, when I was lying in bed, I heard a dramatic act of rape. Opposite the Habsburg hotel, there is Ernest Street.[38] It has amazingly good acoustics, possibly thanks to the broken windows or the holes made by shells in the walls of the buildings. Enough to hear distinctly two voices: a male one, first urging, then cursing, and a female one, first begging "Hilfe", and then just screaming with pain with all its might.[39]

What makes this description so utterly dramatic is that the author is reporting not simply what he can see, but also what he can hear. It is also worth noting that his earlier description of the raped girl led along by her father might give the impression that for him it was the sense of sight that testified to the reality of an incident. Furthermore, Jochelson is clearly distancing himself; his appallingly indifferent tone may indicate the narrative strategy that Karin Bijsterveld terms "adaptation", growing accustomed to the tragedies that were so ubiquitous in the city at that time.[40]

36 Ibid., 140.

37 Ibid.

38 Jochelson probably meant Ernst Strasse, now Tadeusza Rejtana Street.

39 Jochelson, *Kronika*, 213–214. German alarm signals warning of imminent danger and women's cries for help are a recurrent theme in post-war accounts. The theme is widely explored in this book by Sławomir Wieczorek.

40 Wrocław and the western territories (also called the Regained Lands) that Poland was assigned after the war had a bad reputation for being dangerous places, where

Over time, Jochelson could see a steady process of normalisation of living in the city, expressed in the gradual disappearance of intrusive sounds. The once common gunfire is less and less frequent:

> Shooting can only be heard from time to time. More frequent during my first days in Wrocław, in recent days it is more seldom. However, there are some small SS units hiding in cellars and shooting at Soviet or Polish soldiers. I heard the last exchange of fire on St Antoni St. on the eve of my departure.[41]

The author made this note just before leaving Wrocław on 29 June. After being out of the city for quite a long time, on 28 July he resumed writing to give an account of the latest developments:

> Wrocław has definitely changed its appearance. It has become quiet and calm, only the grinding bicycle wheels, still often without tyres, remind one of the fierce times two months ago. The barricades have mostly been removed, and even if some have been left, there are no doubt ways around them. The streets have been swept clean of the red brick dust that before would often reach up to one's ankle, and shooting is now a rare thing, usually a sign of the merriment of a drunken militiaman or soldier. The Germans have become meeker, there are fewer disputes with the Soviet command, the looters with carts have disappeared altogether, and those carrying loot in their hands are few and far between.[42]

Indeed, there is no mention of shooting anymore. At the end of the chronicle, the author remembers single shots again, but they are "sensational" sounds, evidence of fun, not fighting – it was probably "a soused individual" who was shooting.[43]

Wrocław certainly did not become completely quiet, devoid of its characteristic sounds. As late as 7 October 1945, he writes:

---

all kinds of fugitives from justice headed. Joanna Konopińska's journal provides an apt illustration of this perception: "They say in central Poland that the western territories are to be settled by the underclass, all kinds of loafers and thieves, the so-called 'underworld' that the other regions want to get rid of. Maryla [the writer's sister] brought us a newspaper with an ordinance of the National Town Council of Cracow, in which they stated openly who would be displaced to the western territories. So many of the dregs of society have already come here, but it is outrageous that they are making a penal colony of the new lands. Actually, in central Poland especially, Lower Silesia is named 'the Wild West' or 'Mexico', etc., and, sadly, the terms are justified. Perhaps Maryla is right in saying that we should run away from here. Maybe one can't live here" (Konopińska, *Tamten wrocławski*, 77).

41 Jochelson, *Kronika*, 230.

42 Ibid., 234.

43 Ibid., 315.

> The robbery in Rosenthal[44] has surpassed the looting of the previous day. The soldiers did not care to interrupt it; they even took part in the stealing themselves. Having come to know the area very well, they operated there all the more eagerly and proficiently. They would get into a house through holes in the ground floor and then proceed from one house to another through their cellars. At about 1 p.m., some representatives of the TZP turned up:[45] they looked around, then loaded their three-tonne car with all kinds of goods and went back.
>
> At five o'clock, patrols of the Soviet command of the city appeared. No one but they managed to bring some order. But at night, the raping and looting started again. And along with it acts of vandalism: chopping up doors, window frames and furniture with axes, breaking windowpanes.[46]

Yet on the same day, he wrote about the effects of the Soviet military patrols to the districts of Karłowice and Kowale: "the nights have become quiet, not like Wrocław, devoid of gunfire".[47] His notes reveal here how the soundscape changed depending on the time of day and night and the particular district.

With the passage of time, Jochelson began to notice other sounds, not so dramatic, which seemed to signal rebirth. When his wife was out of the city for twelve days, on 26 September, he wrote:

> During her absence, a lot of things have happened: in the city, in the office, and in the house as well. The streets are flourishing with new shops; in almost every building there is a grocery. The tram runs from the end of Wilhelmshafernstrasse[48] to the corner of Stalina St. (former Matthiasstrasse, now Jedności Narodowej St.) and Michaelisstrasse.[49] The market has been moved to the biggest square in Europe, which happened to be named "Grunwaldzki". The opera ensemble staged *Halka*[50] at the theatre. The audience did not like the performance, mainly because Platówna,[51] who played Halka, was too old. Some concerts were held too. However, I stayed at home in mourning.[52]

And some time later, now in 1946, he remarked about children singing:

> The children have got used to Wrocław. They know a lot of poems and songs, they sing: "Płynie Odra, płynie po polskiej krainie. Zobaczyła Wrocław, pewno go nie minie".[53]

---

44 Now Różanka – a district in the northern part of Wroclaw.

45 TZP – Provisional State Authorities.

46 Jochelson, *Kronika*, 278.

47 Ibid., 280.

48 Jochelson probably meant Wilhmeshafener Strasse, now Olszewskiego Street.

49 Now Nowowiejska Street.

50 A Polish national opera by Stanisław Moniuszko.

51 Franciszka Platówna-Rotter (1894–1974), a Polish opera singer.

52 Jochelson, *Kronika*, 267.

53 Ibid., 308. The song roughly translates thus: "The Odra River flows across the Polish land. It has noticed Wrocław, surely it won't pass by".

Joanna Konopińska also took note of some cultural events, the premieres of *Halka* (8 September 1945) and *The Barber of Seville* (6 December 1945), the first film show (16 June 1945), and the opening of the Bristol patisserie and café-dancing hall with a select jazz band (4 November 1945).[54] She observed that *The Barber of Seville* was played by an orchestra composed of German musicians directed by Stefan Barański, a conductor from Cracow, and *Halka* was performed by German dancers, the choir of the Silesian Opera and artists from Cracow.[55]

Another example of what could be heard in the cultural life of the city in 1945 can be found in an excerpt from a recollection of Jochelson published by Włodzimierz Suleja:

> There was a phenomenon, now non-existent, that was a remnant of the occupation. During friendly gatherings, people would declaim patriotic poems, with sincere pathos and often good performance, while playing national music on the piano or the violin.[56]

German compositions must have been heard in the city as well; Konopińska writes that she decided to bring her pre-war piano scores from her family home in Panienka, near Jarocin, because "my teacher only has the German ones with military marches and folk songs".[57]

Polonisation

The drama of the place in which Jochelson is trying to make a home and a living for himself and for others is emphasised not only by the sounds of gunfire, devastation and looting; one can also hear the crying and *panting* of those displaced. The author describes two such moments, the first on 11 August 1945, the other on 7 October the same year:

> In the afternoon, I called a public meeting of the Germans in Rosenthal. There were several hundred people. Kapkayev and I read out the protocol of the "conference of the three [great powers]" in Potsdam, and then we disabused them of the remaining hopes that they evidently cherished. The gathering reacted intensely. When the lieutenant, for example, declared that the whole German nation was guilty, we could hardly stop the

54 Konopińska, *Tamten wrocławski*, 75.

55 Ibid., 107.

56 Jochelson, "Życie kulturalne", 123. Janusz Majer also writes about partisan songs being sung after dinner and piano concerts held at home (Majer, "Akademicka straż uniwersytecka", 164).

57 Konopińska, *Tamten wrocławski*, 142.

> commotion. When I explained that they were facing displacement [...] and that it would be Poland here, I saw some of them cry.[58]

> I paraded on my motorcycle alongside a kilometre-long snaking line of people condemned to repatriation. Handcarts, bedding, old people of both sexes, women and children, all panting from exhaustion, doggedly pushing or pulling... A disgusting sight... the whole road littered with dumped bundles, some also on handcarts. And the procession is followed by a bunch of vultures – Soviet soldiers and [Polish] civilians – throwing themselves greedily on their property. Everywhere there is wailing and lamentation. Somebody fainted at the railway station. Mayor Wachniewski himself was watching that odd parade on the corner of Psiepolska Street and the so-called Querallee.[59]

It is worth emphasising that the "wailing" and "lamentation" cited in the above account are intrusive, unwanted sounds, which most often evoke emotional resonance among the earwitnesses, as well as the readers of this kind of message.

During the first year after the war, the German language was noticeably audible and intrusive, making the newcomers' experience of the city even stranger.[60] The Germans were still in the city, and like the Poles they had to grapple with the problems of everyday life. What is more, they sometimes behaved as if nothing had changed. At the beginning of his life in the city, on 10 June 1945, Jochelson mentions a startling, inconceivable event:

> Today, I saw an interesting picture on Św. Krzyża St. As I was walking down to the cathedral, from a third-floor window of building number 38, I heard a duet. Two people were singing: a German woman leaning her head, with its crop of red hair, on the chest of a German staff officer, possibly a general, who was accompanying her. The officer was wearing full uniform, with decorations, as if nothing had happened, as if Germany hadn't been defeated and "Heil Hitler" was still in force. I could clearly see the bullion epaulette faced in red, so he couldn't have been a doctor, a pharmacist or a veterinary, because they all have purple facing.[61]

One of the tasks on the Polonisation agenda was to fill the city with the sounds of the Polish language. Perhaps that is why Jochelson, who after coming to Wrocław

---

58 Jochelson, *Kronika*, 253. Jochelson does not hear anything, but he can see people cry.

59 Ibid., 275. Querallee – now Poprzeczna Avenue.

60 Thum refers to a significant quotation from Tadeusz Budziński: "The first days in the new place. From the window, one could see and hear children walking with their nannies and singing songs – in German. That language was still heard everywhere. My parents were in two minds about staying or going somewhere 'in Poland'. I witnessed such discussions" (Thum, *Obce miasto*, 215).

61 Jochelson, *Kronika*, 210.

attended Masses in German, reported on his first service heard in Polish. The sounds of Polish are synonymous with the return of the city to "the motherland":

> On 1 November, I drove Reverend Chancellor Figura, appointed coadjutor for the Polish population of Osobowice,[62] to his first Mass in the new parish. It was a most solemn occasion: in the church, in which nobody had ever heard a Polish sound (except in the confessional, because the former parish priest, Revd Ober, a German, used to bring along Polish confessors for his Polish flock), the Mass was celebrated with a sermon in the mother tongue.[63]

In the memoirs of the first mayor of Wrocław, Bolesław Drobner,[64] there is mention of a short article on a thanksgiving service that was celebrated in Polish on 13 May 1945, after which a crowd of people sang "Boże coś Polskę"[65] "in a general spirit of exultation".[66]

On 15 April 1946, in the penultimate entry in her journal, recalling the start of her new life in the city, Joanna Konopińska wrote: "During the first period, the presence of the Germans was noticeable [...] one could hear German in the streets, on the trams and in the shops, which reminded us of the war and all its atrocities".[67] A few lines below, in the same entry, she continued:

> Now that the Polish population is growing and most of Germans have left the city, day by day Wrocław is becoming more likeable and homelike. I have come to like the city – I can't imagine I could ever leave it. Here I have my home, a job, my studies and a lot of new friends. I don't know how it has happened that Wrocław is so close to my heart.[68]

During the years after the war, various local dialects of Polish could be heard in Wrocław. The sound of a dialect helped to identify where someone had come from. Konopińska wrote in her journal a short account of the 1946 New Year Mass:

> The priest was speaking in a soft Vilnius accent, the German organist was playing the tunes of songs I didn't know, and the congregation was singing Christmas carols, everybody in a slightly different way. A woman sitting on my right was singing with a Lviv drawl, behind me a Silesian was articulating the words of the carols in a harsh, thick voice.[69]

---

62 Osobowice is a district in northern Wrocław.

63 Jochelson, *Kronika*, 294.

64 Bolesław Drobner (1883–1968), a chemist and a socialist from Cracow, mayor of Wrocław between April and June 1945.

65 "Boże coś Polskę" [God save Poland], a hymn sung in Polish churches.

66 Drobner, "Zdobyliśmy Polskie", 91.

67 Konopińska, *Tamten wrocławski*, 211.

68 Ibid., 212.

69 Ibid., 139. Julian Cierpisz, who came to Wrocław from Warsaw, responding to the 1966 competition question "What is Wrocław for you?", wrote about its inhabitants:

Unlike German, Polish was expected, longed-for and noticed. On 15 October 1945, Tadeusz Jarmicki, a former forced labourer displaced to Breslau after the defeat of the Warsaw Uprising, opened a bookshop, and when he obtained "the first, the most ordinary Polish reading primer, [he] knew that Wrocław was Polish, would remain Polish, and that new life was beginning to germinate on the German ruins".[70] The sound and the internal resonance of the Polish language (experienced, for example, when Polish signboards were read out[71]) were evidently among the *sensational* and *comforting* sound topoi.

Places

Wrocławians gave accounts of places that were characterised by a special soundscape. Władysława Pilak remembers Oporów, from which came the sounds of gunfire and yelling,[72] while Konopińska mentions Karłowice, a peaceful place where she took her walks.[73] She also remembers her visit to a cemetery on All Saints Day in 1945: "The graves are overgrown with weeds and rose bushes that have run wild, and some have caved in. The paths can hardly be seen in the undergrowth, and there's absolute silence, silence rumbling in the ears".[74] On her way back, she was followed by the sound of "rats roaming the streets in hordes".[75]

The city was vibrant, especially on the street market places. One of them was the famous "szaberplac" ("looters' square") – Grunwaldzki Square. On 1 December 1945, Konopińska wrote:

> There is a lot to see here! People display their goods on the ground: porcelain, underwear, clothes, shoes, cutlery, paintings, carpets, toys, thousands of different things. Most of the trade is done by German women, but not only. In the evenings, the square is deserted, heaps of rubbish are left and nobody clears it away. The next day, other traders come, and

---

"Instead of the artful dodgers of Warsaw [...] a conglomeration of different guys, various dialects, sayings, accents, habits and miscellaneous regional customs". Cierpisz, "Z wielkomiejskością", 250.

70 Jarmicki, "W kanionie ognia", 170.

71 In the entry of 31 October 1945, Konopińska noted: "I was amazed at the opening of a genuine department store, located in the centre of the city on Schweidnitzer Strasse [now Swidnicka St.]. It is a commercial establishment named Śląski Dom Handlowy [Silesian department store]. So we have the first Polish signboards in Wrocław". Konopińska, *Tamten wrocławski*, 64.

72 Pilak, "Kocham w nim", 90.

73 Konopińska, *Tamten wrocławski*, 56.

74 Ibid., 65.

75 Ibid., 65.

> the trading goes on. Countertrade is prevalent. Money has no value. Among the moving crowds, one can't see the militiamen. People are forcing their way through, shouting and stealing; fights are common. The neighbouring streets are empty and dead, only the looters' square is flourishing.[76]

An important role in the life of the new residents was also played by the places where broadcasts were played over loudspeakers. Konopińska was greatly impressed by what she heard on 25 November 1945:

> Today for the first time I heard an announcement made over the megaphone of Wrocław broadcasting centre. I was just coming back from the library, my bag packed with books, when the sounds of music started coming from a megaphone set up in the Market Square, and then a male voice followed with a bulletin and announcements. Passers-by were stopping and raising their heads as they listened attentively to the words streaming from the loudspeaker. [...] Now we have makeshift radio in Wrocław! Ten megaphones have been installed in the busiest parts of the city. They will be broadcasting twice a day: at 1 p.m. and 6 p.m.[77]

Conclusion

The above-quoted descriptions of the soundscape of Wrocław during the first year after the war perfectly render the special ambience that is so well known from historical studies and analyses of accounts which were not orientated towards the sounds of everyday life. Most of the records feature the topos of intrusive sounds, which are used to evoke the difficult experience of a lack of security. The recalled unwanted sounds can also be classified as sound marks of a city in a state of war and anomie, and some of them, today, as its sonic icons. Particularly significant here are sounds made by people in distress, such as crying, wailing and calling for help. Indicative of suffering, such sounds create a strong emotional resonance. Like the narrative strategy of the juxtaposition of contrasting repertoires of sounds, they seem to have a powerful impact on the recipient's imagination. All the accounts, especially of the first days after arrival, contain hardly any comforting or sensational sounds. For the new inhabitants of Wrocław, it was most important to have a sense of security, to feel at home and to "return to normality". The experience of the first sought-after sounds of the city of their hopes and expectations is articulated in their descriptions of the sound marks of

76 Ibid., 103. A month later, she added: "in the early autumn, one could see here mostly German women trading various goods for food. Now the whole area is occupied [...] by both Germans and Poles". Ibid., 149.

77 Ibid., 97.

Poland, the Polish language, Polish music and recurrent information about the gradual disappearance of the unwanted sounds. The mentions of sounds that the authors are familiar with, denoting their national and ethnic identity, signalling the appearance or the need of a soundscape that is consistent with their identity, can be identified as another narrative strategy: "takeover". This strategy does not involve a process of adaptation, but aims at changing the existing sonic environment, subjugating it or at least discovering its similarity with the imagined, homegrown soundscape, consistent with one's own identity.

tr. Krzysztof Tańczuk

Bibliography

Aalbers, Jasper, Bijsterveld, Karin, Ficker, Andreas, and Jacobs, Annelies, "Shifting sounds. Textualization and dramatization of urban soundscape", in Karin Bijsterveld (ed.), *Soundscapes of the Urban Past. Staged Sound as Media Cultural Heritage* (Bielefeld: Transcript, 2013), 31–66.

Bijsterveld, Karin, Introduction to Bijsterveld (ed.), *Soundscapes of the Urban Past. Staged Sound as Media Cultural Heritage* (Bielefeld: Transcript, 2013), 11–28.

– "Beyond echoic memory: Introduction to the special issue on auditory history", *The Public Historian*, 37/4 (2015), 7–13.

Birdsall, Carolyn, "Earwitnessing: Sound memories of the Nazi period", in Karin Bijsterveld and Jose van Dijck (eds), *Sound Souvenirs: Audio Technologies, Memory and Cultural Practices* (Amsterdam: Amsterdam University Press, 2009), 169–181.

– "Sound memory: A critical concept for researching memories of conflict and war", in Danielle Drozdzewski, Sarah de Nardi and Emma Waterton (eds), *Memory, Place and Identity. Commemoration and remembrance of war and conflict* (London: Routledge, 2016), 111–129.

Cierpisz, Julian, "Z wielkomiejskością jeszcze poczekamy…" [The big city will have to wait a while…], in Bohdan Jałowiecki (ed.), *Związani z miastem… Opracowania i fragmenty wypowiedzi nadesłanych na konkurs: Czym jest dla ciebie miasto Wrocław* [Tied to the city… Edited extracts from entries submitted to the competition "What the city of Wrocław means to you"] (Wrocław: Ossolineum, 1970), 249–258.

Damczykowa Felicyta, "Znak 'P' nosiłam jak sztandar" [I wore the "P" sign as a flag], in Anna Kosmulska (ed.), *Niewolnicy Breslau. Wolni we Wrocławiu. Wspomnienia Polaków z wojennego Wrocławia* [Prisoners of Breslau. Free in Wrocław. Memories of Poles from wartime Wrocław] (Wrocław: Wydawnictwo TMW "Wratislavia", 1995), 79–98.

Drobner, Bolesław, "Zdobyliśmy Polskie Złote Runo" [We captured the Polish Golden Fleece], in Mieczysław Markowski (ed.), *Trudne dni: Wrocław 1945 we wspomnieniach pionierów* [Difficult days: Wrocław 1945 in the recollections of the pioneers] (Wrocław: Ossolineum, 1960), 75–106.

Dulczewski, Zygmunt, and Kwilecki, Andrzej (eds), *Pamiętniki osadników Ziem Odzyskanych* [Memoirs of settlers of the Recovered Territories] (Poznań: Wydawnictwo Poznańskie, 1963).

Gleiss, Horst (ed.), *Breslauer Apokalypse 1945: Dokumentarchronik vom Todeskampf und Untergang einer deutschen Stadt und Festung am Ende des Zweiten Weltkrieges*, 10 vols (Wedel: Natura et Patria, 1986–1997).

Hargreaves, Richard, *Hitler's Final Fortress. Breslau 1945* (Mechanicsburg, PA: Stackpole Books, 2015).

Hartung, Hugo, *Śląsk 1944/1945* [Silesia 1944/1945] (Wrocław: Via Nova, 2009).

Jałowiecki, Bohdan (ed.), *Związani z miastem… Opracowania i fragmenty wypowiedzi nadesłanych na konkurs: "Czym jest dla ciebie miasto Wrocław"* [Tied to the city… Edited extracts from entries submitted to the competition "What the city of Wrocław means to you"] (Wrocław: Ossolineum, 1970).

Jarmicki, Tadeusz, "W kanionie ognia" [In a canyon of fire], in Anna Kosmulska (ed.), *Niewolnicy Breslau. Wolni we Wrocławiu. Wspomnienia Polaków z wojennego Wrocławia* [Prisoners of Breslau. Free in Wrocław. Memories of Poles from wartime Wrocław] (Wrocław: Wydawnictwo TMW "Wratislavia", 1995), 160–170.

Jochelson, Andrzej, *Kronika Semipałatyńsk-Wrocław* [Semipalatinsk-Wrocław chronicle] (Wrocław: Towarzystwo Przyjaciół Polonistyki Wrocławskiej, 1997).

- "Życie kulturalne w 1945 r. w polskim Wrocławiu" [Cultural life in 1945 in Polish Wrocław], in Włodzimierz Suleja (ed.), *Wspomnienia wrocławskich pionierów* [Recollections of the Wrocław pioneers] (Wrocław: Poligraf, 1995), 120–123.

Klatko, Stanisław, "Kupił nas Niemiec z Famo-Werke" [We were bought by a German from Famo-Werke], in Anna Kosmulska (ed.), *Niewolnicy Breslau. Wolni we Wrocławiu. Wspomnienia Polaków z wojennego Wrocławia* [Prisoners of Breslau. Free in Wrocław. Memories of Poles from wartime Wrocław] (Wrocław: Wydawnictwo TMW "Wratislavia", 1995), 150–157.

Konopińska, Joanna, *Tamten wrocławski rok 1945–1946. Dziennik* [That Wrocław year 1945–1946. A journal] (Wrocław: Wydawnictwo Dolnośląskie, 1987).

Kosmulska, Anna (ed.), *Niewolnicy Breslau. Wolni we Wrocławiu. Wspomnienia Polaków z wojennego Wrocławia* [Prisoners of Breslau. Free in Wrocław. Memories of Poles from wartime Wrocław] (Wrocław: Wydawnictwo TMW "Wratislavia", 1995).

Kulczyński, Stanisław, "Grupa kulturalno-naukowa" [The cultural-scientific group], in Mieczysław Markowski (ed.), *Trudne dni: Wrocław 1945 we wspomnieniach pionierów* [Difficult days: Wrocław 1945 in the recollections of the pioneers] (Wrocław: Ossolineum, 1960), 107–133.

Kuligowski, Kazimierz, "W ruinach" [In the ruins], in Mieczysław Markowski (ed.), *Trudne dni: Wrocław 1945 we wspomnieniach pionierów* [Difficult days: Wrocław 1945 in the recollections of the pioneers] (Wrocław: Ossolineum, 1960), 134–157.

Kuzior, Jerzy, "Moje pierwsze wrocławskie lata" [My first years in Wrocław], in Anna Kosmulska (ed.), *Niewolnicy Breslau. Wolni we Wrocławiu. Wspomnienia Polaków z wojennego Wrocławia* [Prisoners of Breslau. Free in Wrocław. Memories of Poles from wartime Wrocław] (Wrocław: Wydawnictwo TMW "Wratislavia", 1995), 72–78.

Majer, Janusz, "Akademicka straż uniwersytecka. Posterunek Oporowo" [The academic university guard. Oporowo post], in Włodzimierz Suleja (ed.), *Wspomnienia wrocławskich pionierów* [Memories of the Wrocław pioneers] (Wrocław: Poligraf, 1995), 158–166.

Majewski, Ryszard, *Wrocław-godzina-zero* [Wrocław hour-zero] (Wrocław: Krajowa Agencja Wydawnicza, 1985).

Markowski, Mieczysław (ed.), *Trudne dni: Wrocław 1945 we wspomnieniach pionierów* [Difficult days: Wrocław 1945 in the recollections of the pioneers], vol. 1 (Wrocław: Ossolineum, 1960).

Pilak, Władysława, "Kocham w nim wszystko, nawet gruzy" [I love everything about it, even the rubble], in Zygmunt Dulczewski and Andrzej Kwilecki (eds), *Pamiętniki osadników Ziem Odzyskanych* [Memoirs of settlers of the Recovered Territories] (Poznań: Wydawnictwo Poznańskie, 1963), 81–95.

Siwicka, Irena, "Szpital dla 'Ausländerów'" [The hospital for "foreigners"], in Anna Kosmulska (ed.), *Niewolnicy Breslau. Wolni we Wrocławiu. Wspomnienia Polaków z wojennego Wrocławia* [Prisoners of Breslau. Free in Wrocław. Memories of Poles from wartime Wrocław] (Wrocław: Wydawnictwo TMW "Wratislavia", 1995), 254–272.

Suleja, Włodzimierz (ed.), *Wspomnienia wrocławskich pionierów* [Memories of the Wrocław pioneers] (Wrocław: Poligraf, 1995).

Thum, Gregor, *Obce miasto. Wrocław 1945 i potem* [A foreign city. Wrocław 1945 and after] (Wrocław: Via Nova, 2005).

Urbańska, Wanda, "Ukradziono mi dzieciństwo" [My childhood was stolen], in Anna Kosmulska (ed.), *Niewolnicy Breslau. Wolni we Wrocławiu. Wspomnienia Polaków z wojennego Wrocławia* [Prisoners of Breslau. Free in Wrocław.

Memories of Poles from wartime Wrocław] (Wrocław: Wydawnictwo TMW "Wratislavia", 1995), 132–135.

Zaborowski, Henryk, "Zamiast Białogard – napisano Wrocław" [Instead of Białogard, they wrote Wrocław], in Bohdan Jałowiecki (ed.), *Związani z miastem… Opracowania i fragmenty wypowiedzi nadesłanych na konkurs: Czym jest dla ciebie miasto Wrocław* [Tied to the city… Edited extracts from entries submitted to the competition "What the city of Wrocław means to you"] (Wrocław: Ossolineum, 1970), 175–183.

Sławomir Wieczorek

Institute of Musicology, University of Wrocław

Calls for help and the sounds of pot-banging in the soundscape of ruined Wrocław in 1945

Abstract: This chapter deals with cries for help and the sounds of banging on pots and metal gongs heard in Wrocław in 1945 during the first several months after the fall of Festung Breslau. These sound signals were produced by the German inhabitants of the city as a response to attacks on their homes. In the first part of this text, I analyse the autobiographical accounts of earwitnesses; in the second, I study the narrative strategies applied to the presentation of the sounds in question in Jacek Inglot's novel *Wypędzony* [Exiled].

Keywords: soundscape, sound signals, Wrocław, Breslau, 1945, war, postwar

The sonic environment of Wrocław changed suddenly and dramatically over the twelve months of 1945, giving rise to a vast number of highly diversified sound phenomena. What I focus on in this article is one selected phenomenon that appeared in the ruined city at night during the first few months after the fall of Festung Breslau: calls for help in the form of the sound of various metal household utensils (mostly cooking pots) being banged, accompanied by cries of "Hilfe!", produced by the German population, which faced assault in their own homes. I have traced mentions of this phenomenon in about a dozen published autobiographical accounts: memoirs, diaries and video interviews. Most of their authors are Poles who began coming to the city from May 1945 onward, but there are also German speakers who found themselves in Wrocław during the siege. Analysis of these recollections of pot-banging and cries for help, as well as the thoughts and actions which they provoked, provides a unique insight into the situation of Germans in Wrocław and into the highly varied behaviour of the Polish newcomers toward the Germans. The study of accounts also sheds light on the complexities of Polish memories about the Wrocław of 1945. Włodzimierz Suleja stresses that until 1989 "topics considered as sensitive or [...] indecent would be left out", which also applies to the robberies committed on the German inhabitants by the Poles and the Russians. He conjectures that writers were "afraid that showing compassion to the recent enemy would be simply improper".[1] The

1 Suleja, "Przedmowa", 5–6.

political transformations of 1989 brought an end to the communist propaganda and censorship, but the abolition of official restrictions did not automatically and immediately change the discourse found in descriptions of the city's history. In the accounts which are the subject of my research, comments on the urban soundscape become vehicles of memory concerning the widespread violence against the Germans. The problem has been tackled in various ways depending on the time when the given account was written – from brief and rather veiled remarks on the clangour of metal pots and cries to longer descriptions specifying the source of those sounds. To date, these urban sounds have not been the subject of research, although they were mentioned by the English journalist Richard Hargreaves in his 1945 account of the "land of the dead" (as he calls Wrocław after the end of the military campaign). He wrote: "night after night, the streets resounded to cries for help or the hellish sound of cooking pot lids being banged to drive the plunderers away."[2] These sounds and cries do feature, however, in Jacek Inglot's 2012 novel *Wypędzony. Breslau – Wrocław 1945* [Exiled: Breslau – Wrocław 1945], to which I devote a separate section of my study.

Calls for help and the sounds of pot-banging in autobiographical accounts

The sounds referred to most frequently in the accounts are those produced by banging tin pots, lids, saucepans and other kitchen utensils with metal objects such as hammers. In some texts, we also encounter information about gongs made of pieces of rail. Mentions of the pot-banging, gongs and cries for help appear already in memories of the Polish newcomers' first nights after their arrival in Wrocław. Stanisław Święch wrote about nights in early May 1945, just a few days after the city's surrender: "These were very turbulent nights, and often you couldn't sleep, despite the fatigue. What I remember best is the pitch-black city, the cries, the banging of tin pots and the shooting…"[3] Most of the reports I have gained access to relate these sounds to the summer of 1945. Zdzisław Samsonowicz, who came to Wrocław in July, suggests that the gongs were also used during the battle for Breslau: "Already during the war, during the siege, the Germans organised themselves in such a way that whenever danger came – of any kind, like an air raid – they had gongs, in the form of pieces of hanging rail, a cardboard box or something."[4] On another occasion, he recalled that those gongs remained in the

2 Hargreaves, *Hitler's Final Fortress*, 225.

3 Święch, "Ludzie", 82.

4 Zdzisław Samsonowicz, in Świętnicki, *Mój pierwszy dzień*.

streets and yards of Wrocław even after the war ended, so that they could still serve the German families in moments of threat.[5] Many of the accounts stress the regularity of this phenomenon with phrases like "night after night".[6] According to Bolesław Broś, the nocturnal noises "became an integral part of nights in Wrocław for a long time."[7] The cries and banging were heard in many different parts of the city – both the ruined high-density housing in the centre and the residential districts such as Oporów (Opperau), Zalesie (Leerbeutel) and Zacisze (Wilhelmsruh), as well as areas with mixed types of housing, such as Sępolno (Zimpel). According to R. Murray Schafer's classification of soundscapes, these were sound signals belonging to the category of "acoustic warning devices", along with bells and sirens.[8] In *Acoustic Communication*, Barry Truax points to two features of sound signals which are particularly relevant within the context of my study of the 1945 Wrocław soundscape. He claims that sound signals reflect the character of the community that uses them and can be preserved in the long-term memory of the witnesses.[9]

I do not have any direct testimony from the inhabitants of Wrocław as to who used those signals, while earwitness reports contain few relevant details. Consequently, of key significance are two accounts from the district of Oporów concerning women's voices crying "Hilfe!"[10] and women who had been advised to bang on metal rails in the event of danger.[11] These accounts tally with the findings of historians, according to which, after the fall of Festung Breslau, the majority of the Germans who remained in Wrocław were the elderly, women and children, who could not defend themselves against robberies.[12] Calling for help gave them a chance of rescue. Their German neighbours would only respond to an alarm signal. As a result, the signal spread in the urban space, as confirmed by earwitnesses, such as Hendrik Verton ("the signal warned the neighbours, every one of whom joined in with a clash of cymbals, the din of which reached a crescendo, stretching from house to house"[13]) and Wilhelm Kempfi ("[the Germans] sounded the alarm by banging on all kinds of vessels and crying 'Hilfe, Hilfe'. When such

5 See Samsonowicz, *Wspomnienia*, 34.
6 Majer, "Republika", 292.
7 Broś, "Wielki exodus", 167.
8 Schafer, *The Soundscape*, 10.
9 Truax, *Acoustic Communication*, 58–61.
10 Turczyński, *Rancho*, 371.
11 Andrzej Pachlewski, quoted in Bigda, *Osiedla*.
12 Cf. Ordyłowski, *Życie codzienne*, 13.
13 Verton, *In the Fire of the Eastern Front*, 207.

cries were heard at one end of a district, the whole district would frequently take it up.")[14]

Memoirists described the sound of pot-banging and gongs as elements of "self-defence",[15] "a system of defence"[16] and a "communication system".[17] In the context of the pot-banging, one Polish earwitness wrote about "the hope of receiving help",[18] and that hope was far from vain. Several accounts directly show that the signals were part of a rational and well-planned strategy of self-defence. The signals were received by the Soviet military and Polish armed units (such as the state militia and the academic guard) operating in Wrocław. "The calling out of street names and house numbers in the pitch-black night", an extremely important detail recalled by Ursula Waage,[19] accompanied by banging on metal objects and cries of "Hilfe!", was a way (specific to Wrocław in 1945) of summoning those services that could effectively intervene and help the attacked Germans. This custom also provides evidence of the emergence in the city of temporary, supranational acoustic communities.[20] Therefore Richard Hargreaves's claim that the pot-banging was intended to frighten away intruders ought to be supplemented: that was in fact one of the effects, but not the main purpose of such behaviour.

Both Hendrik Verton and Ursula Waage emphasise that the sounds were addressed to the Soviet troops stationed in the vicinity. Verton mentions "jeeps and soldiers" sent by the Soviet headquarters in reaction to the alarm. The soldiers "enjoyed getting to grips with the Poles, between whom there was no love lost [...] [and] showed no mercy, and very often shots were to be heard."[21] Waage placed her whole account of the pot-banging within the context of information concerning the Soviet command moving out of Sępolno. She wrote in her diary that until recently the alarms had resulted in "interventions from Soviet soldiers", because the Soviet military defended the German civilians from assault and theft. But now, after the Russians had moved their command, "no one will come to our rescue, so the looting raids are becoming more and more audacious."[22] Verton noted that the Germans being frequently rescued by the Russians may apparently be hard

14 Kempfi, *Między Katowicami*, 100.
15 Ibid.
16 Młotkowski, *Młodzież*, 155.
17 Verton, *In the Fire of the Eastern Front*, 207.
18 Broś, "Wielki exodus", 167.
19 Waage, *Przeżyj!*, 68.
20 Truax, *Acoustic Communication*, 57–58.
21 Verton, *In the Fire of the Eastern Front*, 207.
22 Waage, *Przeżyj!*, 68.

to understand, while Gregor Thum writes about an "unsettling friendship" established between the Germans and the Russians soon after the city's capitulation.[23]

The signals were also received by the Polish academic guard, as we read in accounts by its members. The guard was hastily formed, poorly armed and poorly trained. Its task was to guard the property of the university then being established in Wrocław and the safety of the Polish academics moving to Wrocław. In the memories of the members of an academic guard unit from Oporów, we read about German inhabitants who asked them for protection against robbers and against the Red Army unit stationed nearby. The soldiers, as Janusz Majer explains, "used to visit the Germans living on the housing estate in search of alcohol, snacks and women."[24] We also learn from his account that the gongs were an agreed alarm signal: "we called all the locals living at the very end and in the middle of each street to our station. We ordered them to hang pieces of rail out of the first floor windows or the attic, and to raise the alarm whenever necessary."[25] It is notable how the location of the alarm gongs was precisely planned, taking the acoustic effect into consideration. The memories of another member of the Oporów academic guard confirm and supplement the above-quoted account: "German women approached us with requests for help. We advised them to hang pieces of rail over the windows. At moments of danger (the assaults and robberies usually took place at night), they could rouse us by banging on the rails with a hammer, which also informed us where help was immediately needed."[26] The reports of the guard members also tell us about their reactions to the alarms and about the fact that warning shots were an effective defence tool. "When we heard such a gong and the cries of 'Hilfe – help!', we would run there full pelt, shooting into the air. In many cases, this scared the attackers away, but we did not always get to the place on time. And this was repeated night after night."[27] Zdzisław Samsonowicz, who served on the guard in another

23 According to Thum, the Germans established official contacts with the Soviet military authorities, as well as making unofficial arrangements with the soldiers in the hope of securing their help as well as permission to ignore the orders of the Polish administration. See Thum, *Uprooted*, 36–43.

24 Majer, "Republika", 292.

25 Ibid.

26 Pachlewski, quoted in Bigda, *Osiedla*.

27 Majer, "Republika", 292. On the meaning of the warning signals and the alarming silence when they ceased, Eugeniusz Młotkowski wrote: "all the guards in the city soon found a common language, which consisted in shooting into the air to scare assailants away. In the evening, it seemed like fighting was still going on; and when no one was shooting, people became worried because of the silence. In the event of an alarm, we shot twice." Młotkowski, *Młodzież*, 155.

district of Wrocław, describes the reaction of his unit to the signals: "We took the rifles and, if it was not too far, we went there to help in some [way]. Whether they were Germans or not, we were to there to protect people."[28] Interventions of the Polish militia (police) are mentioned in the memories of Wilhelm Kempfi, who lived in the exclusive residential district of Zacisze. Kempfi noted that the banging on metal objects and the cries for help were "in many cases effective", and a militia patrol car would come to their rescue. On one occasion, he adds, the arrival of Polish militiamen resulted in a skirmish and bloodshed between the militia and Soviet soldiers. After that, the militia no longer intervened.[29]

Memories of the academic guard members inform us what kind of action was taken in reaction to the pot-banging signals. Earwitness accounts also include passages which point to the reception of the sounds and the significance attached to them. The reception varied widely, but we can distinguish several characteristic attitudes: from feeling moved or intrigued to irony. I use these epithets for purely descriptive purposes, without passing any value judgements. Sympathetic reactions appear in the testimonies of women: Ursula Waage, who was German, and Lidia Arczyński, who came from a Polish-German family. In 1945, both were young girls directly in danger of being assaulted. In her diary, Waage mentions "hellish noise": "But at night it gets really bad. Sometimes whole buildings are searched by Polish looters. People use pot lids and other kitchenware to make a hellish noise."[30] Lidia Arczyński recalls cries for help: "those voices crying 'Hilfe, Hilfe' were horrible; after the bombings, those cries were the worst thing. They came in the night and woke you up."[31] This comparison of the deafening sound of air raids which had brought death and destruction – by far the loudest sounds heard in Wrocław during the first half of 1945 – with the cries for help is particularly revealing with regard to the horror of the German inhabitants' lives after the fall of Festung Breslau. In accounts given by the Poles, we meet with a different type of reaction.[32] Władysław Kowalczyk describes the banging on lids and pots as

28 Zdzisław Samsonowicz, in Świętnicki, *Mój pierwszy dzień*.

29 Kempfi, *Między Katowicami*, 100.

30 Waage, *Przeżyj!*, 68.

31 Lidia Arczyński, in Świętnicki, *Mój pierwszy dzień*.

32 In this context, it is worth quoting the account of Tadeusz Wronka, although it comes not from Wrocław but from Lubań (Lauban), about 150 km from Wrocław. Wronka describes the cries he heard in the night – both "Hilfe!" in German and "pomocy!" in Polish – as "inhuman cries" which, along with the "sounds of shooting, of doors and windows battered down" became, in his opinion, "the standard sounds of that 'life' we had then". He also adds an important remark about his own attitude to the events

"a rather peculiar acoustic phenomenon of that time".[33] Eugeniusz Młotkowski, a member of the academic guard, initially derides in his memories what he considers the Germans' hysterical behaviour, and he treats the sound signals anecdotally: "Someone only had to knock a bit more loudly on the locked gates, and you got the whole street beating and banging all kinds of gongs and yelling 'Hilfe'."[34]

The quoted passages demonstrate the obvious difference between the accounts of Polish- and German-language witnesses concerning the nationality of the assailants and the defenders of the German population. Ursula Waage and Hendrik Verton unambiguously identify the assailants as Polish looters, while they look to Soviet soldiers for protection. In most memoirs of persons who came to Wrocław in 1945, we read about gangs of Polish looters or thieves, but it is the Soviet soldiers who are mostly held responsible for assaults on the German inhabitants. Two exceptions to this rule are the accounts written by Tadeusz Turczyński and Władysław Kowalczyk before 1989. They write about unidentified perpetrators of assaults, murders, robberies and rapes. The two accounts differ, however, in the way they present the sound signals. Kowalczyk describes unspecified "cries for help and banging on pot lids."[35] Turczyński, on the other hand, makes things quite clear, albeit indirectly. He uses his comments on the Wrocław soundscape to shed light on the anarchy that prevailed in the city and points to the Germans as its victims. In his recollections of service in the academic guard in Oporów, he writes about the "sound of a gong suddenly heard in the night or at dusk" and "women's loud voices crying 'Hilfe!'"[36] He thus identifies both the nationality and the sex of the victims, and he hints at the scale of the phenomenon when he notes that he had already become familiar with these sounds in other districts of Wrocław.

Turczyński returns to the soundscape of Wrocław in one more passage, where the sounds of gong-beating and shooting mix with those of a music performance. He recalls his first evening in the city, when his companion played Chopin on the piano, "most frequently the Polonaise in A flat major [Op. 53]." This music is commonly recognised as an emblem of Polishness, and Turczyński adds that "the

signalled by those sounds: "Who would have felt up to intervening in such things, in that tangle of ruins and darkness, and with the awareness that there was as yet practically no authority in the city? All kinds of accounts were settled in the night – both the old ones from the time of the German occupation [of Poland] and the new ones resulting from the sharing of booty and debates over vodka." Wronka, *Z ruin i zgliszcz*, 323.

33 Kowalczyk, *Wspomnienia*, 203.

34 Młotkowski, *Młodzież*, 155.

35 Kowalczyk, *Wspomnienia*, 203.

36 Turczyński, *Rancho*, 371.

twilight intensified the mood; the voices died down." But rather than continuing in the solemnly patriotic vein, he attracts attention to the sounds heard by those gathered in the room, which remind them of the violence being perpetrated in the city: "single shots can be heard outside the window, and the faraway beating of a gong…"[37] The juxtaposition of the sinister, intrusive sounds of the city with the sounds of music can also be found in the memories of Bolesław Broś. He described a summer evening on the porch of a former German villa. A machine gun usually stood in the corner. This time, those present were listening to "German hits" on a gramophone found in the flat. The context was not safe, however, since – as Broś recalls – "a peculiar nocturnal serenade began" at dusk,[38] consisting of cries coming from the neighbouring houses: pot-banging, grenade explosions and shooting in the nearby Szczytnicki Park (Scheitinger Park). What strikes the reader of Broś's memoir is the ironic use of the name of a musical genre originally performed in the open air (for example, in front of a beloved's window) on a pleasant evening in good weather.

The use of pots and improvised gongs as sound signals can be seen as a symptom of the de-urbanisation and ruralisation of Wrocław – processes analysed with reference to post-war history by Gregor Thum.[39] He claims that the destruction of the buildings turned the city into "an agglomeration of loosely connected villages set in an expanse of rubble."[40] Another reason for these processes occurring, he argues, was the mass influx into the city of the rural population, which maintained its former lifestyle in the new location, for instance keeping farm animals in the city. That is why the voices of farm animals heard in the various districts of the city have so far been considered a key acoustic indicator of those processes. Wrocław had lost its urban character also because of the types of sound signals used for communication in the city. The telephone network had been destroyed during the war, and the German inhabitants had no access to automobiles; the services responsible for maintaining order and security in the city did not work in a regular and comprehensive manner. The sounding of alarms using household utensils made Wrocław like a village. As the inhabitants themselves recall in their written memories, it even became a kind of jungle or forest. Hendrik Verton claims that the "communication system" applied in Wrocław was also used in the jungle, except that instead of tam-tams the city population used "saucepan

37 Ibid., 367–368.

38 Broś, "Wielki exodus", 167.

39 See Thum, *Uprooted*, 98–104.

40 Ibid., 104.

lids".[41] Tadeusz Turczyński points out another similarity: the way sound spreads in space. On the gong-beating that he heard in Oporów, he writes: "it wafts and spreads like the sound of a tam-tam in the forest."[42] The same alarm system was very frequently used by villagers during the Second World War to warn against Ukrainian nationalists attacking Polish villages. In one account, we read: "Every homestead had a gong, a metal bar beaten with a hammer, a piece of sheet metal or some other metal object to raise an alarm with."[43] Another author mentions "banging on aluminium bars and pots hung from the branches of fruit trees."[44] It seems that the repertory of signals used by Poles in those regions to raise the alarm was even more extensive than in the ruined Wrocław. As we learn from witness memories, church bells were also frequently used for this purpose, and during that period they were an alarm signal "by day and by night."[45] The German inhabitants of Wrocław, however, had been deprived of access to even those few church bells that survived the wartime confiscations and were not destroyed during the siege.

The depictions of cries for help and the sounds of pot-banging in Inglot's novel *Exiled*

Written in 2008–2011 by Jacek Inglot, the novel *Exiled. Breslau-Wrocław 1945* is the most important artistic example of interest in the sound signals that are the subject of this article.[46] Those sound signals constitute an important element of the auditory sphere of the novel, recurring several times.[47] The author took his

---

41 Verton, *In the Fire of the Eastern Front*, 207.

42 Turczyński, *Rancho*, 363.

43 Gawęda-Kuchta, *Wspomnienia*, 240.

44 Peretiatkowicz, *Wołyńska samoobrona*, 61.

45 Kwiatkowski, *Życiorys*, 361.

46 Due to its subject, the novel attracted attention mainly in the local media in Wrocław. Several reviewers generally praised the author in the cultural press for the very fact of taking up this topic and going beyond the one-sided national point of view on the city's past, as well as his successful presentation of the realities of 1945, based on historical sources. At the same time, reviewers criticised the weakness of the plot and flaws in the portrayal of the main characters, as well as the stylistic ineptitude. See Hnidiuk, "'Nie była niczym osłonięta'", 23; Pięciak, "Realistycznie", 29.

47 Inglot's novel is not the only book by a Polish author set in Breslau/Wrocław which takes up the motif of pot-banging. In Marek Krajewski's popular crime story *Festung Breslau*, this motif is introduced in a different context altogether: that of a kind of acoustic propaganda war between the defenders of Festung Breslau and the attacking Red Army. At one point, the protagonist notices in the street a group of inhabitants

information about these signals from the memoirs of Wrocław's first Polish inhabitants, on which, as he admits in the commentary included in his book, he drew abundantly while writing his novel.[48] In the afterword, he also points to the ethical intentions which motivated him to take up this subject: "In my novel, I tackle a painful and sensitive issue, which we merely skirted around for many decades now [...] Breslau did not cease to exist in one day, on 6 May 1945, when the fortress surrendered. It was dying slowly for many months afterwards – and it was us Poles who let it die."[49] The novel deals with the tragic fate of the German inhabitants and the complex responsibility that Poles bear for this moment in the city's history. This intention is different from the one that inspired the autobiographical accounts I have quoted and analysed above. The question is, what in this context is the role of the cries for help and the pot-banging? My reading of the novel will focus on an analysis of the *narrative strategies* applied to the sounds in question.

The action of the novel begins in June 1945, when a young Polish underground army fighter, Jan Korzycki, comes to Wrocław from Warsaw to flee persecution by the communist security service. Unexpectedly, he soon joins the Polish (state) militia, and thus becomes a witness to the anarchy prevailing in the city and to the tragedy of the German inhabitants. In the novel, the moment of his arrival in Wrocław coincides with the introduction of one of the main and most frequently used narrative strategies for the presentation of an urban soundscape. Researchers dealing with the textualisation and dramatisation of urban soundscapes call this strategy "arriving in the city". The protagonist's arrival in a city previously unknown to him or her, they claim, is a conventional moment used in films, radio

---

producing "a horrible metallic cacophony" by banging on pot lids (Krajewski, *Festung Breslau,* 176). This sound is a spontaneous attempt to drown out the "roar" of the dance music coming from the Russian loudhailers. Many reports of the siege actually recall the loudspeakers installed in the city by the Red Army. They were mostly used to deliver messages that encouraged German soldiers to desert, but also to broadcast universally familiar musical repertoire. The aim was to weaken the determination of the defenders by playing music that evoked emotions and memories radically contrasting with the current situation. Cf. Hargreaves, *Hitler's Final Fortress*, 154–155. What attracts attention in Krajewski's book is the interesting reversal of the typical reception of sound signals in the wartime context: the inhabitants of the besieged city consciously make noise to drown out the music which would normally be close to their hearts.

48 The author most likely used the account by Tadeusz Turczyński, who called the gongs "tam-tams." In the latter's account, the voices crying "Hilfe!" are exclusively female, and the sounds come through the window from a certain distance. All these elements can be found in Inglot's novel as well.

49 Inglot, *Wypędzony*, 378–379.

plays and written texts to present the city's soundscape.[50] Sound depiction thus reflects the unique character of the new setting of the protagonist's life. Already during his first night in Wrocław, the hero of Inglot's novel hears ("intruding on his sleep", as the narrator tells us) "shooting, pot-banging, and women's desperate voices crying 'Hilfe.'"[51] The quoted passage does not tell us who was hitting the pots or why, but the intrusive character of these sounds is evident, since they are coupled with other sounds pointing to violence and anarchy. It is also emphasised that the victims are German women.

This opening scene (its three key elements) becomes a model for the way this motif is represented throughout the novel. In this and other sections, it is juxtaposed with other sounds; the protagonist always finds himself at some distance from the sources of those sounds and remains passive in the face of the sounds. Let me begin with the juxtaposition of sounds. The very next day, soon after waking up, Korzycki notices that – as a result of the military campaign – the city is totally devoid of birdsong. "He realised that he had neither seen nor heard any birds here. Even if they had survived the siege, they would hardly have anything to feed on."[52] The peculiar character of Wrocław's sound environment, which the protagonist discovers during the first days after his arrival, is reflected in the pot-banging and the cries for help – but also in the lack of birds. This contrast recurs several times in the novel as a key concept for the construction of its sound environment. The author himself wrote about this contrast in his popular science article on Wrocław's life in 1945, published after the release of his novel. Among the phenomena characteristic of "year zero in Wrocław", he mentions the city's peculiar soundscape: "There were no nightingales singing at night in Wrocław. Instead, one could hear shooting, pot-banging and German women desperately crying 'Hilfe!'"[53] This narrative strategy is referred to as a *juxtaposition* of soundscapes. Researchers distinguish two kinds of juxtaposition: in time (a soundscape of the present with soundscapes of the past or the future) and in space (the soundscapes of various parts of the same city).[54] In Inglot's novel, the juxtaposition involves a contrast between the current abnormal post-war landscape and the traditional soundscape of some indefinite time of peace. The sonic components of this juxtaposition are presented side by side, and in nearly every case the other soundscape appears as an absent alternative. For instance, in another passage where

---

50 Bijsterveld, et al., "Shifting sounds", 35.

51 Inglot, *Wypędzony*, 80.

52 Ibid., 85.

53 Inglot, "Wrocław", 27.

54 See Bijsterveld, et al., "Shifting sounds", 49–54.

this juxtaposition occurs, Korzycki commands a group of militia men whose task it is to stop looters plundering whole quarters of the city at night. Waiting for the intruders, he listens attentively to the sounds of the city and "again realises that there are no birds. Killed or scared away during the siege, they would not return. Perhaps one day, when grass overgrows the ruins and the eternal shooting ends..."[55] Suddenly, looters attack the homes of German inhabitants, who respond to the danger by using pots and saucepans to make noise.

As the novel progresses, the narrator notes that the pot-banging gradually becomes rare, and birds return to the city; this is signalled in two phases. One morning, the protagonist is woken up by the song of a nightingale, which "began to sing at daybreak. Birds were coming back. He should probably feel glad, but they had retired very late and now the nightingale's trills were stopping him from sleeping."[56] Soon, however, he hears the nocturnal sounds of the city again, "just the same as usual – shooting, cries and pot-banging",[57] which reminds everyone that Wrocław is still a dangerous city. Gradually, as the protagonist observes the nocturnal soundscape, he first notes the disappearance of the sound of metal objects being struck: "There had been no shooting for quite a while; nor could one hear the Germans making the noise with which they once greeted the looters. He could not recall such a quiet night in Breslau before."[58] This is followed by the final comment on the birds' return to the city: "unexpectedly, they heard nightingales' trills coming from the street. The birds were back for good."[59] The reversal of the sequence of sound phenomena over the course of the narration suggests the restoration of normality and the return of the acoustic order familiar from the times of peace. Interestingly, this kind of normalisation process does not occur in the other aspects of the represented world.

On the basis of an analysis of the testimony of the first Polish settlers conducted by Renata Tańczuk, we should note the quite numerous mentions of birds singing in the city soon after the fall of the fortress (Festung). Tańczuk also points to a narrative strategy present in the memoirs, which consists in juxtaposing birdsong with the sounds of post-war terror: shooting and explosions. There is a crucial difference here: in the accounts of Polish settlers, the two phenomena occur simultaneously; in the novel, they replace each other as alternatives. The best example of this is a passage from Andrzej Jochelson's chronicle quoted by Renata Tańczuk,

55 Inglot, *Wypędzony*, 130.
56 Ibid., 211.
57 Ibid., 221.
58 Ibid., 300.
59 Ibid., 308.

where the author writes about "the traditional sounds of Wrocław [...] shooting, the rumble of mines being blown up by sappers, and in between – the song of nightingales, the cries of swallows, and the warbling of sparrows."[60] Tańczuk draws an important conclusion concerning the immense potential of the juxtaposing of intrusive with comforting sounds for breathing drama into the representations of city life.[61] Inglot's stereotypical "alternation" variant is probably less effective.

Another narrative strategy applied in Inglot's novel is one that I will refer to as "listening from a distance". This term incorporates both spatial and emotional distance. In every instance, the sounds produced by the German inhabitants in times of danger reach the protagonist from afar. The distance may be longer or shorter, but he never finds himself in direct proximity and never sees the people banging on pots and crying "Hilfe!" Most frequently, he hears those sounds through the open window of the room he is in. Even in the last of the quoted passages, Korzycki first opens the window and only then realises that the acoustic indicators of violence are gone. His separation from the sources of the sounds also enhances the contrast between the relative safety of his own situation and the robberies and assaults taking place at the same time in other parts of the city. This is also supposed to explain his attitude toward those sounds, evident in the scene of his conversation with the German professor: "The deep silence was only disturbed by the rhythmical ticking of the clock on the wall [...] and by the nocturnal sounds of the city, which were the same as usual: shooting, cries and pot-banging. They did not have to worry about them, because they came from afar, from another district."[62] Distance from the source of sound leads to indifference (though it could just as well lead to a sense of helplessness).

The reception of these sounds remains unchanged even when the distance from those producing the signals is shorter. In the novel, the distance is diminished twice. In each case, the type and source of the danger that made the Germans use the sound signal is clearly defined. In one instance, a tenement is attacked and a fire results. In the other, there is a robbery on a train. In the former scene, which is the best developed and most interestingly constructed of all the occurrences of this motif in the book, Korzycki and a group of militia men listen from a distance as a tenement house inhabited by Germans falls under attack. "First there was one person banging on pots. A moment later, other metal tam-tams joined in, accompanied by desperate female cries of 'Hilfe, Hilfe'. More and more voices

60 Jochelson, *Kronika*, 209.
61 See Renata Tańczuk's chapter in the present volume.
62 Inglot, *Wypędzony*, 221.

joined those crying, becoming louder and louder, until a minute later the whole district was yelling like mad, banging on pots and saucepans with everything they could lay their hands on. This racket intensified, with new musicians joining in, until shots were fired." When the attack led to a fire breaking out in the house, they cried 'Es brennt! Es brennt!' and banged on the pot lids."[63] In this passage, one is struck by how the author highlights the temporal, dynamic aspect of the sounds by showing how the noise intensifies when more and more people join in, banging on household utensils. The protagonist comments with indifference: "'The chaps have got down to it', muttered Korzycki, snuggling down in the hole." One of his companions reacts differently, however, and asks about the reasons for their passiveness: "'And we what? Do nothing about it? People are being robbed there!' 'And we do nothing',"[64] replies Korzycki, before explaining that the looters are more numerous and better armed, which means that they can only be defeated if a trap is set for them. He does not refer in any way to the threatening sounds they heard, though just a few moments later he betrays that he is sensitive to them when his somatic reaction to the sounds of murder is described: "A cry was heard. Rather a howl than a cry – inhuman, beastly, as if someone was being skinned alive. There was nothing in this cry but monstrous pain. It was piercing, like the scraping of a knife on glass, and it chilled one to the bone. Korzycki felt the skin on his face go numb and his mouth fill with a hideous metallic taste."[65]

The sounds of pot and lid banging and cries for help are presented in the novel as a spontaneous and not entirely comprehensible reaction of the Germans to their life and property being threatened.[66] The Germans themselves remain nameless. We do not know who they are, although the narrator stresses distinctly that the voices – "desperate", as he always adds – crying "Hilfe!" are those of women. The purpose of the pot-banging routine is not explained in the novel. Are these sounds meant to scare away intruders? Or are they (and the cries) a signal, a call for help? If so, to whom are they addressed? The reader is not provided with answers to these questions, because in the represented world nobody reacts to the

---

63 Ibid., 131–133.

64 Ibid., 132.

65 Ibid.

66 The scene of the robbery on the train in particular makes this kind of impression. Korzycki is on board with Germans being expelled from Lower Silesia. During the attack of the Polish gang, "voices and cries of terror were heard; that desperate 'Hilfe' of women which he knew so well from Breslau/Wrocław, and someone even banged on a pot, to which other carriages also responded, banging on whatever came to hand." Ibid., 365.

banging and the calls. None of the characters in the novel relates to them, and the narrator also remains in principle indifferent toward them. The sounds produce no effect in the novel. They are merely an element of the peculiar soundscape of 1945 Wrocław.

In lieu of a summary

As a researcher and an inhabitant of Wrocław, I am interested not only in the past, but also in the possible future of the sounds discussed in this article. I will present this issue in the form of questions. Can those sounds (as in the process described by R. Murray Schafer and Barry Truax) change their status from sound signals to soundmarks, viewed by the acoustic community of Wrocław as unique and particularly valuable for this community? Does this kind of transformation concern (as in Truax's example) exclusively the sound of historical bells or clocks, which are still physically present in the given sound environment?[67] What can trigger this kind of transformation? Can it be achieved by a musical or literary work, a film, a sound installation, a museum exhibition, a theatrical spectacle, or a documentary about the pot-banging and the cries for help? How, if at all, will Jacek Inglot's novel influence further presentations of this phenomenon? We can also ask about the appropriate way of presenting the nonverbal aspects of the said cries for help in relation to their physical, corporeal aspect – their timbres and rhythm.[68] Can the community of Wrocław recognise as soundmarks sounds that represent the complex, varied and ambiguous attitudes of its members in the past – from providing help, through indifference, to engaging in acts of violence themselves? Does the critical dimension of the phenomenon prevent such a transformation, considering the naturally affirmative character of historical policies? How would our discourse on the history of Wrocław have to change to make such a transformation possible? And to ask a more general question – in the memories of Wrocław in 1945, dominated by visual elements, could the representations of sounds win a place equal to that of photographs of ruins and wreckage?

tr. Tomasz Zymer

---

67 See Truax, *Acoustic Communication*, 59.

68 A cry or a call, as Maria Gołaszewska writes, "carries more meaning than the spoken word – it also commands attention with its intonation, timbre, force, and the type of voice modulation." Gołaszewska, *Estetyka*, 78.

Bibliography

Bigda, Beata, "Osiedla Wrocławia – Oporów, odcinek 1 – powstanie Kolonii Akademickiej" [Wrocław's housing estates: Oporów, episode 1, the founding of the academic settlement], *Słowo wrocławian*, http://www.slowowroclawian.pl/oporow, accessed 12 July 2017.

Bijsterveld, Karin, Jacobs, Annelies, Aalbers, Jasper, and Fickers, Andreas, "Shifting sounds: textualization and dramatization of urban soundscapes", in Karin Bijsterveld (ed.), *Soundscapes of the Urban Past. Staged Sound as Media Cultural Heritage* (Bielefeld: Transcript, 2013).

Broś, Bolesław, "Wielki exodus" [The great exodus], in *Pierwsze lata na Ziemiach Zachodnich* [The first years in the Western Territories] (Wrocław: Oficyna Wydawnicza Atut, 2005).

Gawęda-Kuchta, Barbara, "Wspomnienia z Wołynia" [Memories of life in Volhynia], in Lucyna Kulińska (ed.), *Dzieci Kresów* [Children of the Eastern Frontier], ii (Cracow: Wydawnictwo Jagiellonia, 2006).

Gołaszewska, Maria, *Estetyka pięciu zmysłów* [Aesthetics of the five senses] (Warsaw: Wydawnictwo Naukowe PWN, 1997).

Hargreaves, Richard, *Hitler's Final Fortress. Breslau 1945* (Mechanicsburg: Pen and Sword Books, 2015).

Hnidiuk, Jacek, "Nie była niczym osłonięta, a wisiała na gołym drucie" ["She was naked and hung on bare wire"], *Nowe Książki*, 2013/7.

Inglot, Jacek, *Wypędzony. Breslau–Wrocław 1945* [Exiled: Breslau-Wrocław 1945] (Warsaw: Instytut Wydawniczy Erica, 2012).

– "Wrocław, rok zero" [Wrocław: year zero], *Tygodnik Powszechny*, 2013/19.

Jochelson, Andrzej, *Kronika. Semipałatyńsk-Wrocław* [Semipalatinsk-Wrocław chronicle] (Wrocław: Towarzystwo Przyjaciół Polonistyki Wrocławskiej, 1997).

Kowalczyk, Władysław, "Wspomnienia z działalności PPR na Dolnym Śląsku" [Memories of the work of the Polish Workers' Party in Lower Silesia], in Bronisław Pasierb (ed.), *Tworzyliśmy cząstkę nowej Polski – wspomnienia dolnośląskich działaczy PPR* [We were part of the new Poland: memories of Polish Workers' Party activists in Silesia] (Wrocław: Zakład Narodowy im. Ossolińskich, 1974).

Krajewski, Marek, *Festung Breslau* (Warsaw: Wydawnictwo W.A.B., 2006).

Kwiatkowski, Edward, "Życiorys Wołyniaka" [The biography of a Volhynian], in Lucyna Kulińska (ed.), *Dzieci Kresów* [Children of the Eastern Frontier], ii (Cracow: Wydawnictwo Jagiellonia, 2006).

Majer, Janusz, *Republika Oporowo* [The Oporów Republic], *Kalendarz Wrocławski*, 1995.

Młotkowski, Eugeniusz, "Młodzież we Wrocławiu" [Young people in Wrocław], in Włodzimierz Suleja (ed.), *Wspomnienia wrocławskich pionierów* [Memories of the Wrocław pioneers] (Wrocław: Towarzystwo Miłośników Wrocławia, 2001).

Ordyłowski, Marek, *Życie codzienne we Wrocławiu 1945–1948* [Everyday life in Wrocław 1945–1948] (Wrocław: Zakład Narodowy im. Ossolińskich, 1991).

Peretiatkowicz, Adam, *Wołyńska samoobrona w dorzeczu Horynia* [The Volhynian population's self-defence in the River Horyn basin] (Katowice: nakł. autora, 1997).

Pięciak, Wojciech, "Realistycznie i baśniowo" [Realistic and fabulous], *Tygodnik Powszechny*, 2013/21.

Samsonowicz, Zdzisław, *Wspomnienia o Straży Akademickiej Politechniki we Wrocławiu* [Memories of the Academic Guard of Wrocław Polytechnic] (Wrocław: Oficyna Wydawnicza Politechniki Wrocławskiej, 2002).

Schafer, R. Murray, *The Soundscape: Our Sonic Environment and the Tuning of the World* (Rochester: Destiny Books, 1994).

Suleja, Włodzimierz, "Przedmowa do II wydania" [Preface to the second edition], in Włodzimierz Suleja (ed.), *Wspomnienia wrocławskich pionierów* [Memories of the Wrocław pioneers] (Wrocław: Towarzystwo Miłośników Wrocławia, 2001).

Święch, Stanisław, "Ludzie, zwierzęta, zdrowie" [People, animals and health], in Włodzimierz Suleja (ed.), *Wspomnienia wrocławskich pionierów* [Memories of the Wrocław pioneers] (Wrocław: Towarzystwo Miłośników Wrocławia, 2001).

Świętnicki, Witold, *Mój pierwszy dzień* [My first day], Episode 19, TVP Wrocław (2013).

Thum, Gregor, *Uprooted. How Breslau became Wrocław during the Century of Expulsions*, tr. Tom Lampert and Allison Brown (Princeton: Princeton University Press, 2011).

Truax, Barry, *Acoustic Communication* (Norwood: Ablex Publishing Corporation, 1984).

Turczyński, Tadeusz, "Rancho Oporów" [Oporów Ranch], in Mieczysław Markowski (ed.), *Trudne dni. Wrocław 1945 r. we wspomnieniach pionierów* [The hard days: Wrocław 1945 in the memories of the pioneers], i (Wrocław: Zakład Narodowy im. Ossolińskich, 1960).

Verton, Hendrik C., *In the Fire of the Eastern Front: The Experiences of a Dutch Waffen-SS Volunteer*, tr. Hazel Toon-Thorn (Solihull: Helion and Company, 2007).

Waage, Ursula, *Przeżyj! Zapiski z Festung Breslau i polskiego Wrocławia 1945–1947. Relacja świadka* [Survive! Notes from Festung Breslau and the Polish Wrocław 1945–1947. A witness report], tr. Arkadiusz Barędziak (Wrocław: Oficyna

Wydawnicza Atut, 2016); Ger. orig. *Bleib übrig: Aus den Tagebuchaufzeichnungen in der Festung Breslau und der Nachkriegszeit von Januar 1945 bis April 1947. Zeitzeugenbericht* (Halle: Projekte Verlag, 2004).

Wronka, Tadeusz, "Z ruin i zgliszcz" [From the ruins and the wreckage], in Kazimierz Bidakowski, Maciej Iłowiecki and Tadeusz Wójcik (eds), *Pamiętniki inżynierów* [Memoirs of engineers] (Warsaw: Czytelnik, 1966).

Dorota Błaszczak

Polish Radio Archives,
Fryderyk Chopin University of Music, Warsaw

Waves of Remembrance: Wrocław in Radio Sounds. Broadcasting from the past

Abstract: *Waves of Remembrance: Wrocław in Radio Sounds,* an interactive sound installation, created a listening environment for the public. A prepared radio receiver "broadcast from the past", using content from radio archives. This article presents the installation with reference to the role of radio broadcasting and the experience of listening to radio as two inspirations for creating the work. The war-time context and the radio soundscape of 1945 are emphasised as a turning point in the history of the city, and as such they are present in the radio recordings that shaped the installation.

Keywords: radio art, broadcasting, soundscape, interactivity, perception

In May 2015, the Goethe-Institut issued a call for projects: "As part of Wrocław's tenure as the European Capital of Culture 2016, we are looking for projects in the areas of performance and contemporary art that can be realised in a container (about 6 x 3 x 3 m) with three transparent walls located in a public square in Wrocław (the Goethe-Institut Pop Up Pavilion). […] The pavilion will serve as the basis for a site-specific, interdisciplinary cultural programme with a German-Polish focus, whose aim is to include the public and encourage it to participate".[1]

One of the themes for submissions, referring to "identities and remembrance/ Wrocław's German-Polish history", was relevant to the field I was working in and triggered the idea of an interactive sound installation, *Waves of Remembrance,* based on radio sounds connected with Wrocław.[2] The project was accepted for a two-week exhibition, from 27 April to 9 May 2016, in the Goethe-Institut Pop Up Pavilion

1 When the call was issued, in May 2016, it was presented on the webpage of the Goethe-Institut, which now carries information about the event in Polish and German, https:// www.goethe.de/ins/pl/pl/kul/sup/wro.html, accessed 26 February 2017.

2 *Waves of Remembrance: Wrocław in Radio Sounds* (2016): Dorota Błaszczak – concept and programming, Wojciech Barcikowski – selection of archive recordings, Zbigniew Pietrzak – preparation of the radio receiver;
cooperation and support:
Polish Radio Archives, Wrocław Radio Archives, National Digital Archives,

in Nowy Targ Square (Fig. 1).[3] Visitors could set the channels on a radio receiver with the buttons, use the knobs to tune the radio in time on a special scale from 1924 – the year radio broadcasting was launched in Wrocław – to 2016, listen to various historic recordings about Wrocław or switch to the current local radio station. In this journey in time, many layers of history recorded in sound overlapped, invoked by listeners' interactions with the receiver. The glass pavilion helped create an interesting public context to the privacy of listening to radio and the private soundscape generated inside the space of the pavilion. As Michelle Hilmes wrote about radio: "the diminishment of physical distance and penetration into private space is linked explicitly to the spread of culture – and cultural hierarchies. Radio promised simultaneity of experience without direct contact, exposure to the public in the privacy of one's home".[4]

Figure 1: Waves of Remembrance, *Nowy Targ Square, Wrocław (2016). Photo Dorota Błaszczak*

Goethe-Institut, https://www.goethe.de/ins/pl/pl/ver.cfm?fuseaction=events.detail&event_id=20710975, accessed 26 February 2017,
WRO Art Center, http://wrocenter.pl/en/fale-pamieci/, accessed 26 February 2017.

3 More images can be found on my website, at http://dorotablaszczak.pl/home_/2016/iv-v.2016.html, accessed 26 February 2017.

4 Hilmes, "Radio", 354.

Waves of Remembrance was inspired by and relates to the notion of radio broadcasting as a technical process and an important medium which had an impact on people and the soundscape. The following sections will precede a review of the installation and will cover issues related to broadcasting, listening to radio and radio's influence on the soundscape in relation to the period of the Second World War.

Broadcasting and the war

The invention of radio created the possibility of reaching people over a vast area and broadcasting the same message to various places regardless of distance. "Radio is an alteration of space and a structuring of time. It extends space if you're making music, shrinks it if you're listening".[5] Radio became a medium used for establishing or re-establishing a nation, as in the case of Canada, as a huge, diverse country,[6] or Poland, as regaining independence in the 1920s.

The same function of re-establishing a nation can be discerned during the Second World War in response to the fragmentation of Poland, the dispersion of its people and the suppression of the Polish language in the occupied country. Radio broadcasting became an important source of information about the country for people who were far from Poland. A diary from Ankara describes the emptiness of the day when the Warsaw I radio channel stayed silent, 7 September 1939,[7] whilst Fryderyk Jarosy's poem "Warszawa druga" [Warsaw II] reflects the importance of another Warsaw radio channel.[8] These examples show how the sound of Polish broadcasting was an important factor in people's lives during the war.

Another "potential" of broadcasting was used during the war: the potential to influence others. The power of broadcast sound was well understood. From September 1939 to June 1945, possession of a radio receiver in Poland carried the death penalty. From 16 October 1939, the Polish language was banned from radio. A "fake" soundscape was created, with a "fake" German transmitter on the "true" Polish radio frequency, so listeners might think they were listening to a Polish programme.[9] Nazi recordings were made to be broadcast in Poland so

5 Berland, "Toward a creative anachronism", 41.

6 Ibid., 39. "The emergence of a concept of Canada as a nation was dependent on, and articulated through, the building of the national railway and, subsequently, at first literally in its tracks, the national broadcasting system".

7 Kunert, "Cztery pożegnania", 22.

8 Ibid., 23.

9 Kondracki, "Rola warszawskiej rozgłośni", 219.

that the power of the occupier could be heard in statements and music through speakers installed on the streets.

During the Warsaw Uprising, Polish broadcasting became a symbol of hope that Poland still existed. People risked their lives to build radio transmitters, including the "Błyskawica" (Lightning),[10] which functioned from 8 August to 4 October 1944. A few short programmes were broadcast during the day, in both Polish and English, as they were meant for both Warsaw residents and an international audience.[11] After two months, shortly following the last day of the Uprising, the transmitter was destroyed,[12] so it could not be used by the occupiers.

This shows the power of broadcasting, which remained an important tool for rebuilding the country in 1945, just after the end of the Second World War. As an "immaterial" medium, radio was again a means to reach many people and deliver political information. But radio had an important role on a private level, as well, broadcasting personal information to families searching for their relatives. Listening to the radio retained its significance.

Listening to radio: The radio soundscape

Radio sound is a technological construct. It requires the transmission of a source sound. Speakers and amplification are always involved on the side of the receiver. The listener's gesture of tuning into specific radio waves is a necessary part of the process of finding a selected channel. Special attention is required to tune and listen to a radio. The position of the listener in front of a radio set with sound sources hidden inside creates an acousmatic situation for the act of listening.[13] As described by Michel Chion,[14] such a situation can intensify casual listening, when the listener tries to guess the cause of the sound, or, on the contrary, it can help with "reduced listening" to the sound itself. One way or another, listening to radio resembles an acousmatic situation, with the receiver serving as a contemporary

10 Miszczak, *Historia*, 215–217.

11 Nowak-Jeziorański, *Kurier z Warszawy*, 334–335. "Błyskawica" transmissions were received at the radio monitoring centre near London, and a few of the broadcast programmes were stored on records. There are copies of those recordings in the Polish Radio Archives.

12 Kunert, "Cztery pożegnania", 31.

13 "Acousmatic […] marks the perceptive reality of sound as such, as distinguished from the modes of its production and transmission". Schaeffer, "Acousmatics", 77.

14 Chion, *Audio-Vision*, 32.

Pythagoras' curtain,[15] which can help the listener to concentrate on the meaning and emotions conveyed by the radio sound.

It must be pointed out that the original sound source transmitted by a radio station is not the same as the sound received by the listener. Radio sound can be associated with Roman Ingarden's notion of the "musical work".[16] Radio waves are transmitted as intentional objects, and the experience of listening to radio gives rise to sound concretions. This experience depends on the listener's interactions forming a time structure and mixing streams of sound. The actual sound includes the transmission artefacts, superimposed onto the original sound,[17] and the final audio result is determined by the amplification and speakers of the receiver used for listening. The listener's point of view defines the subjective sound, encapsulating the experience of listening – what we could call a radio soundscape.

All types of sounds involved in the process of radio tuning create a specific radio soundscape in response to the listener's interactions with the radio receiver. Such a dynamic soundscape can be analysed using R. Murray Schafer's taxonomy of soundscape features: keynote sounds, signals and soundmarks.[18]

I would consider that a keynote sound, in the case of a radio soundscape, is not a fundamental sound, but a property of that soundscape. The sound of the radio can be present or switched off, so it disappears immediately. The listener can easily switch to another channel and a completely different radio programme; he or she "edits" the soundscape by manipulating the receiver. This property defines a radio soundscape as a technological, not a natural, construct.

Signals as foreground sounds that are listened to consciously can be found in the time signal heard on the radio every hour.[19] Like all other signals, the time signal carries a message for those listeners who can interpret the tones as indicators

15 Schaeffer, "Acousmatics", 77. "In ancient times, the apparatus was a curtain; today, it is the radio and the methods of reproduction, along with the whole set of electro-acoustic transformations, that place us, modern listeners to an invisible voice, under similar conditions".

16 Ingarden, "The sounding", 115.

17 Most of the recordings preserved in audio archives are the original recordings meant for broadcasting, but some were preserved as audio recordings from the receiver side, with all the broadcasting artefacts added, as radio "Błyskawica" programmes from the Warsaw Uprising in 1944.

18 Schafer, *The Soundscape*, 9–10.

19 There is a standard for time signals broadcast on radio, introduced at the dawn of broadcasting, in 1924, at the BBC (Greenwich Time Signal). The time signal with five short pips and one longer beep of 1kHz is still used on Polish Radio.

of the start of an hour. This layer of a radio soundscape requires a semantic type of listening, by means of which the sound may be interpreted.[20]

A soundmark – a unique sound of the radio soundscape – can be heard when tuning a receiver as a static noise when no radio station is present for the frequency of radio waves set on the receiver scale.[21] The static can be heard on the older analogue radios,[22] so this soundmark is related to technology used in the listening process and does not apply to new devices.

Another of Schafer's concepts applies very well to the radio soundscape, namely, the soundwalk. "The soundwalk is an exploration of the soundscape of a given area using a score as a guide. The score consists of a map, drawing the listener's attention to unusual sounds and ambiances to be heard along the way".[23] The scale of the radio receiver, filled with many locations, constitutes a "score" for a radio soundwalk all over the world: by tuning into different radio stations, the listener can visit the soundscapes of distant places. The installation *Waves of Remembrance* represents an invitation to embark on such a soundwalk in time.

The notion of the radio soundscape relates both to a location and also to a specific time in history, such as a time of war, which is relevant to the 1945 soundscape.

Radio and the soundscape during the war

In this section, I will address three different views on the war soundscape that are related to broadcast sound: the presence of broadcast sound within the general soundscape, the character of radio sound, and the nature of listening to the radio.

During the war, even if radio stations were still broadcasting, their programmes could be heard only when radio receivers were accessible, which was not so obvious. At that time, radio receivers were on the move. Many of them were confiscated in occupied countries and given to people in Germany or to the Germany

---

20 In the sense of the listening modes presented by Michel Chion. Chion, *Audio-Vision*, 25–31.

21 Analogue radio broadcasting is based on the modulation of a carrier signal by an audio input signal that can be heard on the radio receiver. The frequency of a radio station is the frequency of the carrier signal. If there is no carrier signal transmitted – when, for instance, a station has stopped working – the receiver will give a static noise at that point on the radio scale.

22 Radio receivers adjust the gain of the signal that was causing the static. Many radios can automatically mute the receiver when no station is detected, so the static is no longer heard.

23 Schafer, *The Soundscape*, 213.

army, so that propaganda could be delivered to a target audience.[24] During the Warsaw Uprising, in 1944, over one hundred radio sets acquired from the Philips factory were given to the residents of Warsaw for the same reason, so they could listen to important news about the situation in the city.[25] On the other hand, some radio receivers were kept in secret by Polish people to listen to the news, even if possessing a radio and listening to it were forbidden.

This information is confirmed by people who lived through the war. My colleague Wojciech Barcikowski's grandfather, Juliusz Lenczewski, had a hidden radio receiver in the German tobacco warehouse in Tomaszów Mazowiecki where he was working during the war. He could secretly listen to the BBC news in the evenings. My own grandfather, Józef Błaszczak, placed a radio receiver inside a stool at the start of the war so the whole family could listen to the news from London, as my father, Gerard Błaszczak, described. In those two stories, both grandfathers talked about the London radio station using the Polish onomatopoeic phrase "bum, bum, bum, buuum" to describe the anticipated BBC broadcast signal. That was Morse code for the letter "V", for "Victory", and it became part of their long-term sound memory.

The sound of radio appeared mainly for short periods of time during the day, and it could easily disappear for various reasons. People would wait eagerly for radio broadcasts, and it was a big loss when the radio signal was absent and they could only hear static. The radio soundscape was split into broadcasting time and waiting for radio sounds (as we learn from soldiers' diaries).[26] Also relevant were power cuts, which caused the radio to fall silent on the transmission or reception side. So the radio soundscape acquired its "switchable" character on different levels: for purely technical reasons, through a lack of electricity and a lack or malfunctioning of receivers, or for political reasons, with transmissions banned altogether.

Another important element of radio sound is the radio voice, "perceived by the listener as factual and informative, newsworthy, or at least dedicated to the betterment of life",[27] as described by Frances Dyson. The sound qualities of a voice were particularly important during wartime. The presenter's voice needed to show authority, but also to bring hope and be calm.[28]

---

24 Miszczak, *Historia*, 213; Fikus, "Hans Fritzsche", 95.

25 http://sp2put.utp.edu.pl/radioelektronicy/technika_amatorska.htm, accessed 26 February 2017.

26 Wawer, "Samodzielna Brygada", 198.

27 Dyson, "Radio Voice", 167. Dyson discusses the need for different radio voices in general and in radio art, but without focussing on wartime.

28 Kondracki, "Rola Polskiego Radia", 216.

The sound of radio brought news and music into the soundscape, but it could also contain coded messages for military purposes – supporting an air defence system, for example, as happened in Poland at the beginning of the war.

When there was a ban on listening to the radio in general (as in Poland) or listening to foreign radio stations (as in Germany), deciding to listen to the radio could require hiding a radio set and keeping it muffled. On the other hand, if radio was used officially, it became part of the street soundscape in various ways. Receivers could be placed in windows by city residents, as broadcasters encouraged people to do,[29] so the radio was "shared" and more people could hear it. But the sound could also be transmitted through cable radio to permanent speakers installed on the streets, so everybody would be forced to listen to the same programme. The radio soundscape was under the control of propaganda. The usual freedom to select radio channels was suspended.

The radio soundscape of 1945

The end of the war and the changes it brought were spread over time and space as the new Poland took shape. The radio industry and radio stations were destroyed and had to be rebuilt. The first transmitter of the new Polish Radio was launched in Lublin on 11 August 1944,[30] while "Błyskawica" was still working at the beginning of the Warsaw Uprising. The double radio soundscape reflected a complex political situation during the last part of the war.

The beginning of 1945 brought the first radio programmes in Warsaw for street speakers only; then a small radio transmitter started broadcasting on 16 March 1945 and the rebuilt Raszyn radio station on 19 August 1945. The programmes were based on live announcements, news (forty per cent of airtime) and music programmes (thirty-four per cent), including live concerts and a very small set of fifteen records, with Polish music – mostly folk.[31] When broadcasting began after the war, those recordings were repeated several times a day. 1945 was a year of one central radio programme prepared in Warsaw and broadcast between eight and sixteen hours a day.[32]

Already in 1945, radio stations were rebuilt in many cities: Cracow, Katowice, Poznań, Bydgoszcz, Gdańsk and Łódź. The following year, stations were launched

29 Ibid., 213.

30 Miszczak, *Historia*, 227.

31 Ibid., 236–237.

32 Ibid., 331.

in Szczecin and Wrocław.[33] There were cable connections for transmission between radio stations, so a large part of Poland could listen to programmes as soon as people had access to radio receivers and electricity. First, from 1945, cable radio stations were established, to give more people access to radio broadcasting as soon as possible. Then, in 1946, radio receivers began to be built in Poland.[34] So the 1945 radio soundscape was mostly based on receivers left in Poland after the war. To return to my father's war memories, when he returned to Warsaw in February 1945, he dug out a radio receiver that had been hidden in a stable in 1939. Unfortunately, it had stopped working, because of its lengthy exposure to humidity. But then, using crystal detectors and spare parts, my father managed to build a crystal set that did not require electricity to listen through headphones. The main purpose of building this receiver was the upcoming broadcast of a football game between British Army and Polish teams in Warsaw on 10 November 1945.[35]

In 1945, sports transmissions and other radio programmes brought the promise of normal life. The national anthem was played on the radio at the beginning and the end of each day's broadcasting.[36] Several radio orchestras were created and gave concerts in 1945, such as the Polish Radio Symphony Orchestra (March 1945), Poznań Orchestra (May 1945) and Bydgoszcz Orchestra (April 1945). The first live piano concert on the radio was Władysław Szpilman's recital broadcast on 19 February 1945.[37] There were also programmes prepared for listeners in rural areas twice a week, featuring agricultural topics and folk music. Educational programmes started in 1944. The first live sports programme was broadcast on 15 August 1945.[38] Live transmissions of political events began with a military funeral in March 1945. Holy Mass was transmitted from 3 March 1946 to 11 September 1949, when such transmissions were forbidden again for many years.[39]

In 1945, most radio programmes were broadcast live; some were recorded onto discs and magnetic tapes, and some of those recordings were held in archives and gave rise to collections of recordings. The Polish Radio Archives became a catalyst for creating *Waves of Remembrance,* to enable visitors to explore recordings about Wrocław.

---

33 *Waves of Remembrance* includes a recording of the official opening of the Polish radio station in Wrocław (16 November 1947).

34 Ibid., 321–325.

35 A recording of this football game (33-P-139) is held in the National Digital Archives.

36 Miazek, "Audycje", 267.

37 Markowska, "Muzyka", 320.

38 Listowski, "Sport", 434.

39 Górski, "Audycje religijne", 294.

The operating principle behind *Waves of Remembrance*

Waves of Remembrance consisted of a real, but "prepared" radio receiver, a virtual receiver on a screen, two additional speakers and a computer that "broadcast" sounds from the past.

The radio receiver was modified by Zbigniew Pietrzak to serve as an interface for the listeners to access historical sounds broadcast over ninety years of radio history.[40] This prepared radio became a time machine for archive sounds to be replayed over the receiver's speakers. The knobs and buttons were modified to be used by listeners as normal manipulators of the radio, but they acted differently. Instead of changing the parameters of the circuits and antenna, they controlled the computer application. An artificial magic eye added feedback for tuning,[41] with a green light indicating a properly tuned sound station.

Figure 2: A prepared radio with special time scale and buttons with thematic channels

1924 . . . . . 1945 . . . . 1980 2016

Off	Radio	||	People	History	History	Culture/Sport

The key function of the prepared radio was to tune "through time" to a selection of historical recordings, thanks to a special scale on the radio's front glass panel consisting of important dates in radio and Polish history (Fig. 2). The visitors could listen to different channels by switching the real buttons of the receiver: the

---

40 It was a Tatry tube radio receiver, manufactured in 1964 by ZKR (Zakłady Radiowe im. Kasprzaka) in Warsaw.

41 The magic eye tube was a special tube used as a green tuning indicator. It is an element of old radio sets remembered well by radio users, as became clear while talking to the visitors to this and another public installations using a radio set.

first button switched off the radio, the second button directed the listener to the current, live broadcast of Radio Wrocław,[42] and a special "antenna" button played a mixture of sounds from all channels broadcast at the same time. The last four buttons on the right offered a choice of thematic channels about people, history (two channels), culture and sport, like selecting a channel on an ordinary radio.

As an interactive installation, *Waves of Remembrance* required a computer application to run during the exhibition to simulate the broadcasting and receiving of audio recordings in response to visitors' interaction with the radio receiver.[43] The receiver sent data about the manipulators through an Arduino microprocessor to the computer running the application. The application's main algorithm defined the process of choosing sound recordings based on the selected channel and the point on the scale. The key principle behind the algorithm was based on the simulation of broadcasting sound; the recordings were not selected "on-demand". So the listener could only tune into sound files that were currently playing.

The application generated a virtual representation of the radio, with information about the current recording displayed on the monitor, and two streams of sound to be played on the speakers of the receiver itself and on two additional speakers.[44]

A special demo mode was prepared for the night, when the installation was not turned off. It automatically tuned slowly through time, so the labels for the recordings and the light of the magic eye were changing and visible to people passing by the Pavilion.

The audio recordings for *Waves of Remembrance*

The audio recordings were selected by the historian Wojciech Barcikowski, my colleague from the Polish Radio Archives, based on three main sources: the Polish Radio Archives, Wrocław Radio Archives and National Digital Archives.

The first group of recordings concerned the people of Wrocław. It consisted of the recorded voices of the residents of the city, covering the Polish period in the

---

42 There was a modern tuner circuit installed inside the receiver, tuned to Radio Wrocław.

43 The MAX 7 programming environment (https://cycling74.com/) was used to program all the audio functions, and an Arduino microprocessor to connect the computer with the radio interface.

44 An image of the radio with a graphic simulation of the movements of the tuning indicator and short descriptions of the sounds were visible on the screen of a monitor hanging from the ceiling. The speakers placed at the back of the Pavilion, behind the radio, were used for additional layers of sounds related to sounds previously sought and "heard" by the visitor.

history of Wrocław. The voices included people of various professions, such as the composer Ryszard Bukowski, writers Maria Dąbrowska, Stanisław Dygat and Wojciech Żukrowski, theatre director Jerzy Grotowski, zoo director Antoni Gucwiński, musician Lech Janerka and poet Tadeusz Różewicz. The channel playing these recordings was organised in alphabetical order across the whole scale of the receiver.

The other three channels were structured in chronological order across the scale so the listener could tune through time by turning the knob. These channels contained many recordings about various historical, cultural and sporting events from the Weimar Republic and the period of national socialism ("Reichssender Breslau"), from post-war times, from the independent "Radio Solidarność Wrocław" and also from contemporary materials. The visitor might hear recordings about the World Congress of Intellectuals in Defence of Peace (1948), quadruplets from Wrocław (1954), a smallpox epidemic (1963), the Odra 1204 computer (1967), a flood in the city (1997), the decision regarding the hosting of the Euro 2012 football tournament (2007), the European Capital of Culture (2016), and also classical music recordings of the cathedral choir, in Thomas Stoltzer's *O admirable* (1936), the radio symphony orchestra, in Anton Bruckner's Ninth Symphony (1937), Stanislaw Moniuszko's *Halka* at the Wrocław Opera (1961), the Jazz Festival (1974) and the Actors' Song Festival (1988), among others.

The wartime content selected for *Waves of Remembrance* was a mixture of military, political and everyday programmes and included the following audio recordings:[45] Adolf Hitler at the Reichstag (1939), historical monuments of Wrocław (1941), visiting the Polish Pavilion at the Trade Fair in Wrocław (1941), a programme in Hungarian broadcast from Wrocław (1941), the funeral of pilot Werner Moelders (1941), the German Gymnastics Championships (1942), Adolf Hitler after the attempted assassination (1944), cultural news for Silesia (1945), Grand Admiral Karl Doenitz (1945) and the opening of the Polish Radio station in Wrocław (1947).

The selected audio recordings were prepared for playback on the computer application. This included editing and work on the overall loudness balance for all the sound files. In most cases, I decided not to use sound restoration, so the timbre variety and some distortions were kept. Such aesthetic decisions about the

45 The main sources for the recordings from the Reichs-Rundfunk-Gesellschaft, Reichssender Breslau included German radio recordings. There are original records in the National Digital Archives and copies on tape at Polish Radio Warsaw. Several original sounds from the war and radio programmes about the war can be found on the Polish Radio website, at http://www2.polskieradio.pl/wojna, accessed 26 February 2017.

degree of "mastering" of all the recordings allowed for a diversity of the potential soundscapes created during the exhibition.

Waves of Remembrance as an interactive soundscape

The subtitle of *Waves of Remembrance, Wrocław in Radio Sounds,* indicated the provenance of the recordings that could be heard over the old radio set. All the recordings came from radio archives, and they allowed visitors to build up a story about Wrocław and its residents by listening to the receiver.

In my projects, I am interested in exploring the possibility of extending the listening process. This interactive sound installation creates a radio soundscape that can be navigated through time and through the sound streams of the thematic channels. The radio set lets the listener recreate radio soundscapes from different periods of time. The project would not have been possible without the invention of audio recording, enabling ephemeral sounds to be stored, together with all the emotions contained in recordings of voices, which are missing in the text itself.[46]

The installation was not meant to be an augmented reality project; having done it, however, I would consider it as such. Radio as a medium has some important properties that naturally fit into the augmented reality category that is so prominent in modern discourse. Radio sound always needs amplification and speakers, so it is natural to listen to radio augmented reality over speakers (it is not so natural if you try to recreate the sounds of an environment coming from natural sources). In the case of radio, you always tune to a programme that could be live or pre-recorded days or even years before. The listener is always confronted with overlapping time and broadcasting from the past, so the attributes of augmented reality can be found also in listening to a normal radio set. The radio reality and the temporal relationships between recordings are complex and continually augment the listener's reality.

One key aspect of *Waves of Remembrance* is time and sound overlapping, which addresses the history contained in audio archives. This aspect was inspired by digitisation processes occurring in an environment where many concurrent streams of playback sounds were surrounded by the real sounds in the rooms of the archives; the two types of sounds create a random context for each other. This generated the idea for a separate channel enabling a special type of listening to "all" the radio waves that might surround us at a given time. There was a special "antenna" button (originally used for changing the properties of the signal reception), which

46 There is an archive recording from 1954 of the Polish writer Zofia Nałkowska talking about the emotions conveyed by recorded voices compared to the written text alone.

allows the visitor to listen to a mix of all the thematic channels for a particular point on the receiver scale. The listener can use the tuning knob to search for various mixes on the scale. Moving the knob introduces delay-based processing with speed-dependent parameters, so the user might obtain different sound results from manipulating the knob.

The time overlap concept is also realised by assigning a live feed from Radio Wrocław to one of the radio's channel buttons. Furthermore, a time signal is generated every hour, based on the computer system's clock. On top of that, recorded time signals "from the past" can be heard according to their own timeline. The time signal triggers a semantic type of listening to the sound conveying the time information. If the signal from a digitised sound file is heard at a different time than the start of the current hour, the time in the soundscape being created is unexpectedly relativised.

There is another stream of sound in *Waves of Remembrance,* an extension to the listening modes, in the form of an audio mix built up from sounds "collected" by listening. If the visitor is just listening to one radio programme, without moving the knob and buttons, this additional sound stream develops slowly over time. It is heard from the additional speakers only, which need to be placed higher and behind the radio set, as if those "already heard" sounds were entering a cloud of sound memory, shared with others through the act of listening.

If the installation is used with the five main thematic channels set by pressing the radio buttons, *Waves of Remembrance* follows the "archive strategy" from the taxonomy of interactive art proposed by Ryszard W. Kluszczyński: "the dominant factor of the archive strategy is information, (audio)visual data gathered, organised and made available to the audience".[47] If the installation is used with the special "antenna" button set, the receiver may be turned into an instrument: this mode of functionality follows the "instrument strategy", in which the interactive experience is of a "purely performing" character: "It is organised around an interface, which plays a fundamental role".[48]

During the opening, one visitor made everyone worried about the radio by playing with the buttons and knobs, but most people were very gentle while searching and tuning into different recordings. The exhibition was overseen by attendants who were prepared to provide an explanation about the installation. During the first couple of days, I observed and talked with some visitors and with the attendants. The audience consisted of locals, tourists and various groups. There were students who tried to listen through the whole scale and all the channels, and

47 Kluszczyński, *Sztuka interaktywna,* 236.

48 Ibid., 222.

also visitors who entered the pavilion and listened to other people's choices. I met some tourists from Germany who called their friends to listen together to German historical sounds, which triggered a discussion about the history of our countries. They also raised the question of their collection of photographs from Breslau which they wanted to donate for public use. I noticed also that the radio set itself was bringing back memories even without the sound: people started to talk about their relatives who had a similar radio, describing the room it was in. Sometimes they would check that the receiver was modified and had a "fake" magic eye and would remember playing with its green button, which could be pushed in.

The exhibition lasted for two weeks and attracted nearly 1800 visitors.

Waves of Remembrance, prepared in the Polish Radio Archives, represented the continuation of a series of site-specific projects conceived within my work environment and based on its resources.[49] Such projects follow the "active archiving strategy" proposed by Piotr Krajewski of the WRO Art Center,[50] as "long-term preservation by active use of the content. Active archiving can be divided into two types: developing genuine user interfaces and working with archival materials in order to re-create them".[51] Using the radio receiver as an interface to the audio archives created a special context in which to re-listen to audio recordings as intermingled "waves of remembrance".

Acknowledgments

Waves of Remembrance was supported by the Goethe-Institut, WRO Art Center, Polish Radio Archives, Wrocław Radio Archives and National Digital Archives. My special thanks go to Wojciech Barcikowski, Zbigniew Pietrzak, Waldemar Listowski, Dorota Krakowska, Agnieszka Kubicka-Dzieduszycka, Carolyn Birdsall, Sławomir Wieczorek and Glen Fraser.

---

49 Including *Colours of Archives* (1996), *Sound Drawings* (2014), *Hearing Data. Data Carriers* (2014) and *Radio Receiver of 90 Years* (2015).

50 That strategy was introduced by Piotr Krajewski during a Media Library conference at the Museum of Screen Culture in Moscow, in November 2012.

51 From an English synopsis of the presentation *The WRO Collection: Active Archiving,* delivered by Agnieszka Kubicka-Dzieduszycka and Bartek Korzeniowski to the conference Digitizing Contemporary Art, held on 10 May 2013 at the Academy of Fine Arts and Design in Wrocław.

Bibliography

Berland, Jody, "Contradicting media: Toward a political phenomenology of listening", in Jonathan Sterne (ed.), *The Sound Studies Reader* (New York: Routledge, 2012), 40–47.

– "Toward a creative anachronism: radio, the State and sound government", in Daina Augaitis and Dan Lander (eds), *Radio Rethink. Art, Sound and Transmission* (Banff: Walter Phillips Gallery, 1994), 33–44.

Chion, Michel, *Audio-Vision: Sound on Screen* (New York: Columbia University Press, 1994).

Dyson, Frances, "The genealogy of the radio voice", in Daina Augaitis and Dan Lander (eds), *Radio Rethink. Art, Sound and Transmission* (Banff: Walter Phillips Gallery, 1994), 167–186.

Fikus, Sebastian, "Hans Fritzsche – komentator i sternik radia Trzeciej Rzeszy" [Hans Fritzsche: radio commentator and radio steersman of the Third Reich], in Andrzej Ossibach-Budzyński and Krzysztof Jasiewicz (eds), *Polskie Radio w czasie II wojny światowej* [Polish Radio during the Second World War] (Warsaw: Polskie Radio SA, ASPRA-JR, 2015), 85–105.

Górski, Krzysztof, "Audycje religijne w Polskim Radiu" [Religious programmes on Polish Radio], in Andrzej Ossibach-Budzyński (ed.), *Polskie Radio. Historia-Program-Technika. 90 lat Polskiego Radia* [Polish Radio: history, programmes and techniques. 90 years of Polish Radio] (Warsaw: Polskie Radio SA, ASPRA-JR, 2015), 285–305.

Hilmes, Michelle, "Radio and the imagined community", in Jonathan Sterne (ed.), *The Sound Studies Reader* (New York: Routledge, 2012), 351–362.

Ingarden, Roman, "The sounding and nonsounding elements and moments of a musical work", in *The Work of Music and the Problem of Its Identity*, tr. Adam Czerniawski, ed. Jean G. Harrell (London: Macmillan, 1986), 83–115.

Kluszczyński, Ryszard W., *Sztuka interaktywna. Od dzieła-instrumentu do interaktywnego spektaklu* [Interactive art: from the work as instrument to the interactive spectacle] (Warsaw: Wydawnictwa Akademickie i Profesjonalne, 2010).

– "Strategies of interactive art", *Journal of Aesthetics & Culture*, 2010/2, http://www.aestheticsandculture.net/index.php/jac/article/view/5525/6191, accessed 26 February 2017.

Kondracki, Tadeusz, "Rola warszawskiej rozgłośni Polskiego Radia w obronie stolicy we wrześniu 1939 roku" [The role of the Warsaw station of Polish Radio in defending the capital in September 1939], in Andrzej Ossibach-Budzyński and Krzysztof Jasiewicz (eds), *Polskie Radio w czasie II wojny światowej* [Polish Radio during the Second World War] (Warsaw: Polskie Radio SA, ASPRA-JR, 2015), 207–224.

Kunert, Andrzej Krzysztof, "Cztery pożegnania Polskiego Radia 1939–1945" [Four farewells to Polish Radio 1939–1945], in Andrzej Ossibach-Budzyński and Krzysztof Jasiewicz (eds), *Polskie Radio w czasie II wojny światowej* [Polish Radio during the Second World War] (Warsaw: Polskie Radio SA, ASPRA-JR, 2015), 21–42.

Listowski, Waldemar, "Sport – transmisje – sprawozdawcy" [Sport: broadcasts and reporters], in Andrzej Ossibach-Budzyński (ed.), *Polskie Radio. Historia-Program-Technika. 90 lat Polskiego Radia* [Polish Radio: history, programmes and techniques. 90 years of Polish Radio] (Warsaw: Polskie Radio SA, ASPRA-JR, 2015), 421–450.

Markowska, Elżbieta, "Muzyka na antenie Polskiego Radia" [Music on Polish Radio], in Andrzej Ossibach-Budzyński (ed.), *Polskie Radio. Historia-Program-Technika. 90 lat Polskiego Radia* [Polish Radio: history, programmes and techniques. 90 years of Polish Radio] (Warsaw: Polskie Radio SA, ASPRA-JR, 2015), 307–328.

Miazek, Ryszard, "Audycje Polskiego Radia dla wsi" [Rural programmes on Polish Radio], in Andrzej Ossibach-Budzyński (ed.), *Polskie Radio. Historia-Program-Technika. 90 lat Polskiego Radia* [Polish Radio: history, programmes and techniques. 90 years of Polish Radio] (Warsaw: Polskie Radio SA, ASPRA-JR, 2015), 261–284.

Miszczak, Stanisław, *Historia radiofonii i telewizji w Polsce* [The history of radio and television broadcasting in Poland] (Warsaw: Wydawnictwo Komunikacji i Łączności, 1972).

Nowak-Jeziorański, Jan, *Kurier z Warszawy* [Courier from Warsaw] (Cracow: Znak, 2009).

Schaeffer, Pierre, "Acousmatics", in Christoph Cox and Daniel Warner (eds.), *Audio Culture: Readings in Modern Music* (London: Continuum, 2004), 76–81.

Schafer, R. Murray, *The Soundscape. Our Sonic Environment and the Tuning of the World* (Rochester: Destiny Books, 1994).

Wawer, Zbigniew, "Samodzielna Brygada Strzelców Karpackich. Organizacja, walki, życie kulturalne, polskie radio" [The Polish Independent Carpathian Rifle Brigade: organisation, fighting, cultural life and Polish radio], in Andrzej Ossibach-Budzyński and Krzysztof Jasiewicz (eds), *Polskie Radio w czasie II wojny światowej* [Polish Radio during the Second World War] (Warsaw: Polskie Radio SA, ASPRA-JR, 2015), 191–206.

Daniel Brożek

Wrocław

The soundtrack for the art installation 'Glitter': An attempt to reconstruct the soundscape of a post-war cinema in Wrocław within the context of experimental electronic music

Abstract: The 'Glitter' installation produced by Daniel Brożek and Łukasz Paluch for the group art exhibition *The Germans Did Not Come* was informed by the history of the first post-war cinemas in Wrocław and their repertoire. The soundtrack was based on a form of sound archaeology, including reverberation in the cinema hall, the sound of the projector, and the talking, snoring and coughing of viewers, mixed with excerpts from film soundtracks – a collage of sounds and words synthesising the sounds of the post-war reality, an amalgam of Eastern and Western culture, the sounds of a city that no longer exists. In my essay, I discuss examples of contemporary and experimental music illustrating different approaches to the creation of imaginary sound environments that were never recorded: Gavin Bryars's *The Sinking of the Titanic*, Bolesław Wawrzyn's "I", Alessandro Bosetti's *Notebooks* and Peder Mannerfelt's *The Swedish Congo*.

Keywords: sound art, sound archaeology, sound design, experimental music, cinema

In 2014, I was invited to contribute to the group art exhibition *The Germans Did Not Come*,[1] and to prepare an imaginary soundtrack for the installation 'Glitter'.[2] The main theme of the exhibition was the situation of the Polish city of Wrocław after the Second World War, when all the Poles were expecting that after a while the Germans would return to the city and take back what they had owned before

1 Michał Bieniek (ed.), *The Germans Did Not Come.* For full information about the exhibition: http://muzeumwspolczesne.pl/mww/kalendarium/wystawa/niemcy-nie-przyszli/?lang=en, accessed 22 April 2017.
For the exhibition publication with all artwork descriptions: http://muzeumwspolczesne.pl/mww/wp-content/uploads/2012/06/MuzeumWspolczesneWroclawJednodniowka-wersja_do_druku.pdf, accessed 22 April 2017.

2 Artwork photo documentation: http://www.anomalia.pl/pl/,1?nidh=125, accessed 22 April 2017.

the war. The resultant sense of temporariness was mixed with a move to re-Polonise many aspects of everyday life.[3]

The idea for 'Glitter' was inspired by part of the book *Życie codzienne we Wrocławiu 1945–1948* [Everyday life in Wrocław 1945–1948],[4] which describes the history of the first cinemas in Wrocław after the war and their repertoire. The cinemas, most of which do not exist anymore or have fallen into ruin, were in their heyday at that time. Although the repertoire was dominated by Soviet films, the rare American or English productions broke all box office records, enabling citizens to forget about the trauma of war and experience the splendour of Western culture for a brief moment. In war-ravaged Wrocław, dirty and swarming with criminals, the cinemas offered a semblance of luxury and prosperity.

The first part of the artwork is the inscription 'Glitter', in the form of a banner stylised with modernistic typography. The surface of the banner is riddled with what resembles bullet holes in building walls (Fig. 1). Next to this banner, the exhibition curators wanted to create a room resembling the interior of a Wrocław cinema of the late 40s. The visual side of the installation was quite an obvious choice: a screen, an old projector, seats and a lamp asking for "silence". But the room available for the installation was quite small, and we wanted to create the feeling of a cinema hall. Consequently, the sound design became our tool of choice for transforming the available real space into the desired one. Our idea for the soundscape consisted of echoes of the cinema hall, the noises of the projector, and the chatting, snoring and coughing of the viewers, mixed with fragments of film soundtracks processed with old tape recorder sounds, representing a sort of sound archaeology. Inside the room, visitors could hear sounds from films that the inhabitants of Wrocław might have watched in the 40s, accompanied by Polish Film Chronicle newsreels. What emerges is a collage of sounds and words, a synthesis of the sounds of the post-war reality, an amalgam of Eastern and Western culture – sounds of a city that no longer exists.

3 See Thum, *Uprooted.*

4 See Ordyłowski, *Życie codzienne.*

Figure 1: Photo: Małgorzata Kujda

The history of contemporary and experimental music contains examples of different approaches to the creation of imaginary sound environments which were never recorded. Gavin Bryars's *The Sinking of the Titanic* (Obscure Music, 1975) was based on imaginary sounds of the orchestra playing as the *Titanic* sank. Bryars's inspiration for the work came from a report that the ship's wireless operator Harold Bride had witnessed the house band continue to perform as the *Titanic* went down. In April 1912, Bride had told *The New York Times:* "The band was still playing. I guess all of the band went down. They were playing 'Autumn' then. I swam with all my might. I suppose I was 150 feet away when the *Titanic* on her nose, with her after-quarter sticking straight up in the air, began to settle – slowly [...] the way the band kept playing was a noble thing [...] and the last I saw of the band, when I was floating out in the sea with my lifebelt on, it was still on deck playing 'Autumn'. How they ever did it I cannot imagine. That, and the way Phillips (the senior wireless operator) kept sending after the Captain told him his life was his own, and to look out for himself, are two things that stand out in my mind over all the rest".[5] Bryars imagined that the sound would continue to reverberate as it disappeared under the waves: "the music goes through a number of different

5 Bryars, "The sinking of the Titanic".

states, reflecting an implied slow descent to the ocean bed, which give a range of echo and deflection phenomena, allied to considerable high frequency reduction".[6] The composition was started in 1969 as a piece for the contemporary collective Art & Language. It was initially a single page of A4 paper with instructions as to how the work should sound and how it might be created, but not a score as such. In the world of classical music, that was an innovative approach, focused not on a score and notes, but rather on the sound structure.

A slightly different example of artwork born between sound creation and imaginative music is given by the Polish sound artists known collectively as Bolesław Wawrzyn: "One day, by chance at a railway station [Bolesław Wawrzyn] bought a book on the broad aspects of parapsychology. He read there about the phenomenon of Kirlian photography. According to people who work on that subject, it allows you to register the aura of a living being on a film. Under the influence of that book, he began to wonder about the aura of leaves and trees. What they would tell us, what they would like to share with us, if they could? After some time, armed with a MiniDisc he started to capture sounds of surrounding forests and record nature. Basically, he listened to trees and silence. After some time, the idea evolved and places of executions and suffering were recorded. MiniDisc was replaced by equipment of a slightly better class, and forests gave way to sites where plants and monumental trees standing for many years are often the only remaining witnesses to the atrocities that had occurred there. The witnesses most exact, who saw everything and everybody that was there to be seen…"[7] Bolesław Wawrzyn's releases contain plain field recordings from the sites of military forest graves, concentration camps, mass executions, genocide and war cemeteries. His attempts to record silence are connected with highly emotional and suggestive contexts, representing a perfect example of an imaginative approach to music. They also show the great potential of silence in terms which John Cage described in his *Experimental Music* lecture in 1957: "Hearing sounds which are just sound immediately sets the theorizing mind to theorizing, and the emotions of human beings are continually aroused by encounters with nature. […] Emotion takes place in the person who has it. And sounds, when allowed to be themselves, do not require that those who hear them do so unfeelingly. The opposite is what is meant by response ability".[8]

6 Ibid.

7 Liner notes to the Bolesław Wawrzyn album *I* (BDTA, 2013), https://bdta.bandcamp.com/album/i, accessed 22 April 2017.

8 Cage, *Silence*, 10.

Experimental practice connected with an archaeological approach was used by Italian composer and sound artist Alessandro Bosetti in his award-winning radio play *Notebooks.*[9] This soundwork is based on a series of notebooks written in the years 1904–1928 by the Czech composer, music theorist and folklorist Leoš Janáček. Those notebooks contain words, sounds and melodies heard by Janáček and written in musical notation. This kind of sound photography is popular nowadays with portable records that are cheap and easy to use; at the beginning of the twentieth century, meanwhile, the ephemeral obsession with preserving sound moments of everyday life made Janáček include his daughter's last words before her death.[10] In his work on the notebooks, Alessandro Bosetti has tried to bring back to life what he imagined to be the setting for those Janáček annotations by using his voice and creating electroacoustic tableaux in different rooms at Mies van der Rohe's Villa Tugendhat in Brno, mixed together with words from pages in the notebooks generated by means of a speech synthesis application. Bosetti's approach to the restoration of Janáček's memories resembles Luc Ferrari's concept of *musique anecdotique,* opposed to Pierre Schaeffer's *musique concrète* in terms of the relationship between sounds and the listener's memories.

The digital sound synthesis that Bosetti employs in a direct way to recreate natural sounds and human voices is applied in a different way in the work of Swedish composer Peder Mannerfelt. His record *The Swedish Congo* (Archives Intérieures, 2015[11]) contains music recorded using only electronic synthesisers, with no samples or recordings of real human voices. Yet the music itself resembles the original sounds and voices of traditional tribal Congo music recorded by Belgian filmmaker Armand Denis during the 30s. Modern popular electronic music is full of samples from African music. For Mannerfelt, the decision to avoid samples in his music, instead constructing the sound material from scratch, shows his political attitude: "Simply sampling the original album could be seen as another way of colonising or disrespectful appropriation. However, by re-sculpting the

9 For more on the radio play: http://www.rozhlas.cz/radiocustica/projekt/_zprava/alessandro-bosetti-the-notebooks--1457295, accessed 22 April 2017.
For more on the award: https://www.ebu.ch/contents/news/2015/04/euroradio-alessandro-bosetti-win.html, accessed 22 April 2017.

10 Those words formed the basis for Janáček's composition *Elegy on the Death of Daughter Olga,* JW 4/30.

11 Peder Mannerfelt, *The Swedish Congo Record,* https://archivesinterieures.bandcamp.com/album/the-swedish-congo-record, accessed 22 April 2017.

album, reshaping its original musicality into a wild electronic universe of his own, Mannerfelt pays tribute to the traditional and folkloric meaning of the dances".[12]

I find it interesting to see how far electronic music has come from the early twentieth-century avant-garde, focused on formal and absolute music, to the highly conceptualised and contextualised art forms of the twenty-first century. The examples invoked in this essay show that political awareness and sensitivity to social problems, paired with the use of experimental music practices, offer contemporary artists new possibilities for evolving artistic attitudes. Not only can digital sound design tools and modern composition techniques be used to recreate soundscapes from a time when recording technology was not available, they also offer new interpretations and contexts for historical discourse.

Bibliography

Bieniek, Michał (ed.), *The Germans Did Not Come* (Wrocław: Wrocław Contemporary Museum and Art Transparent Foundation, 2015).

Bryars, Gavin, "The sinking of the Titanic at Xebec", http://www.gavinbryars.com/Pages/titanic_xebec.html, accessed 22 April 2017.

Cage, John, *Silence: Lectures and Writings* (Hanover: Wesleyan University Press, 1961).

Thum, Gregor, *Uprooted. How Breslau became Wrocław during the Century of Expulsions*, tr. Tom Lampert and Allison Brown (Princeton: Princeton University Press, 2011).

Ordyłowski, Marek, *Życie codzienne we Wrocławiu 1945–1948* [Everyday life in Wrocław 1945–1948] (Wrocław: Zakład Narodowy im. Ossolińskich, 1991).

12 Ibid.

Dorian Lange

Aalto University School of Arts, Design and Architecture

The muteness of war-time trauma: A nonverbal perspective on the relationship between trauma and soundscape

Abstract: The soundscape of 1945 is approached here within the context of war trauma. Analysis of the sociopolitical and narrative discourses of the post-war period conducted from this perspective illuminates the obstacles to attempts at verbalising the trauma-affected reality. The limits of verbal approaches to accessing a traumatic past are further investigated from the perspective of post-traumatic stress disorder, showing that there is a need to pay more attention to nonverbal modes of expression. I argue that trauma should be made perceptible beyond what can be discovered from verbalised accounts. This raises the question of the effective means for investigating contexts that are notably characterised by traumatic circumstances and turns attention to artistic concepts and artistic research indicating the potential of using sound as a medium for circumventing the verbalisation barrier.

Keywords: war trauma, nonverbal, post-traumatic stress disorder, soundscape, artistic research, narrative, verbalisation

When initially asked if I would consider contributing to this book, I was instantly reminded of the NATO bombings of Yugoslavia that I experienced in 1999. In a matter of seconds, the publication title *Sounds of War and Peace* short-circuited to my memory of the eleven weeks of almost uninterrupted bombings that demolished not only infrastructure such as power plants, electricity lines, refineries, oil depots, roads, bridges, railways and airports, but also many homes and private properties. The targets even included passenger buses and trains. Hospitals were attacked, and cluster bombs were responsible for many civilian deaths.

I recall the mass media full of unbelievable pictures, and a raging noise penetrating every home every day, over and over, confirming that the pictures were real. Yet it was not just the sound of the explosions; every bomb within a radius of tens of kilometres resembled an earthquake and created an immensely powerful wave of air rippling out from its epicentre in all directions. Even the blind and the deaf knew it was war. Again! As the only solution to the complex political conflicts that the so-called civilised world had to offer. In the middle of Europe, on the threshold of the new millennium.

The days were long and noisy. The howling of the air raid sirens several times a day triggered the worst feelings and dismal thoughts. As a teenager, I could only think of the stories and films about the Second World War that I had read, heard and seen as historical documents of the times when my grandparents were about my age. It made the situation even more incredible, and at times quite surreal. Yet the sound of the aircraft approaching, flying overhead and circling soon after the warning sirens was not the same as the one I knew from the documentaries. Given the technical power behind the aggressive noise of the engines, there was no mistaking that it was indeed 1999, and the world had invested much more in the development of weapons than in practical diplomacy and the forging of peaceful and effective conflict solutions.

So the announcement of the forthcoming conference triggered within me a vivid memory of the evening of 15 April 1999. And although I knew it was only a recollection of a past event, I felt just as if it was happening again: that most dreaded sound we were warned about was approaching. Even without experiencing anything like it before, I understood that at this very moment my days may be numbered. Accompanying the noise of the aeroplane engines leaving the airspace above us was a high-pitched hissing, fizzing and whizzing, which grew exponentially louder, evoking a panic reflex in me that I simply could not rationalise. It was the worst way to demonstrate the intuition of the Doppler Effect of a massive, rapidly approaching object. The seconds of fearing the noise feel endless. And then comes an enormous blast. A massive burst of air shooting through the hallway is pushing me against the doorway. The shattering of window panes after the vast droning caused by tremendous vibrations cannot even be heard anymore. I feel in my torso as if something has detonated inside me.

A family house, just 200 metres from our home, was hit and destroyed in this incident. About two weeks later, on 3 May, in a US Defence Department News Briefing, the use of depleted uranium in the attacks on Yugoslavia was confirmed.

Approaching the muteness of war trauma

I have spoken of this incident several times over the past eighteen years to analysts, friends, colleagues and strangers, but usually only when asked about my experiences or when it seemed relevant and informative for others to hear about them. Trying to verbalise the experience, I always felt incapable of telling the whole or even the true story. Expressed in words, the whole account would become incredibly distant to me, to the extent of appearing ordinary and trivial. It has been rather frustrating to realise that no words can suitably convey the entire experience. Even today, words still cannot adequately represent what it really is that requires a proper form of expression – for me to be able to deal with it and to share it with others. As a result, trying to put this burden into words has never been a rewarding act, and a significant part of my trauma seems to remain mute, beyond all words. This realisation has been the main motivation for my engagement in the topic and the incentive to investigate the foundations and mechanisms behind the verbalisation barrier within the context of war trauma. While verbalised experiences can be articulated in both spoken and written words, even written narratives are sometimes read aloud, repeated and shared with others in oral form, and they rarely stay completely inaudible. So the product of verbalisation is to a great extent also an auditory phenomenon, and the link between the two spheres – the verbalisation issues and the sound – is another aspect that I have been curious about.

The inability to talk about one's own individual traumatic experience is a common reaction or symptom in traumatised persons. However, it is worth noting that not all individuals who experience trauma become severely traumatised or encounter difficulties due to the exposure to life-threatening stress that affect them long term. After all, almost every person is exposed to traumatising circumstances at some point in life. While it is common for every individual to show a strong physical and mental reaction in response to severe shock, some manage to return to normal life after a temporary phase of dealing with the incident, while others either continue their existence in this state for a significantly prolonged period of time or become overwhelmed by their unsettled past at a later stage in their lives. Those who do not suffer from the consequences of their traumatic experiences to the degree of developing long-term difficulties are not considered as traumatised in this sense. That does not necessarily mean that their exposure to the stress was less severe. The likelihood of being affected by trauma to the extent of developing distressing symptoms – a state associated with the clinical concept of post-traumatic stress disorder – is determined by a complex interaction of individual physiological and psychosocial factors. So not everybody faces problems with verbalisation either.

This text approaches the context of the verbalisation barrier from the perspective of individual war trauma. The year 1945 should be thought of here as a figure representing a turning-point within a discourse of severe, widespread, accumulated war traumas. The official end of the Second World War filled that figure with the traumas of persecutions, deportations, mass killings, imprisonment, destruction, displacement, lost identities and annihilated families, building on a background that was fully saturated with the traumatic memories and experiences from the devastation of the First World War just a couple of decades before. Beginning with observations relating to the sociopolitical circumstances before and shortly after the Second World War, I intend to demonstrate that the fact that many kept silent about their individual war-time traumas was an inevitable outcome of the given conditions, and not just a symptom of their traumas. Speaking about trauma is impossible if its narrative is inhibited by either the prevalent public discourse or the individual psychosocial context. Therefore, in the second part of the introductory discourse analysis, the issue of inhibited traumatic narratives is addressed from the perspective of narrative theory. In the next section, I will contextualise the limits of verbal approaches to trauma by referring to the ways in which trauma is addressed in the discourse of post-traumatic stress disorder. A critique of the fixation on verbal modes in trauma-focused cognitive-behavioural therapeutic approaches is linked to the assertion of the necessity to pay more attention to the nonverbal representation of trauma in soundscapes. I refer here to my interpretation of the recent meta-studies and reviews demonstrating the suboptimal effectiveness of the commonly used psychotherapies for post-traumatic stress disorder, elaborated in another article. The critique of the verbal approach to trauma is further supported by findings from neuroscience, which brings in another focus of my investigation, namely, the exploration of the potential for transforming trauma using sound as the operating medium. In the concluding section, nonverbal artistic approaches to working with trauma are brought into focus and considered within the context of the soundscape.

The analysis of the verbalisation problem focuses on connecting verbally inaccessible trauma with soundscape studies. The intention of the text at this point is to explore the possibility of applying ideas from different disciplines to the concepts of considering and working with soundscapes. In so doing, I approach the soundscape as a medium concerned mainly with how traces of war trauma evading the verbal context can be made visible or audible or be otherwise expressed, unlocked and revealed in both theory and artistic production. My sphere of action lies in the cross-disciplinary intersection of trauma studies, sound studies, psychotraumatology, artistic research and art therapy. The text therefore represents an attempt to stimulate trauma-informed soundscape studies through a nonverbal, artistic perspective. It does not do justice to the vast amount of scholarly literature on trauma, nor does it provide a discussion of existing writings that touch on the relationship between trauma, memory and sound(scape). Rather, it questions and elaborates the limits of verbal modes for dealing with trauma in the given context, both in the individual sphere and in public discourses, by presenting arguments collected from different viewpoints.

For a better understanding of the discussion that follows, it is worth making an introductory remark on the concept of the soundscape as I define it in this work. A soundscape is understood as a sum or a collection of all audible elements, that is, perceivable acoustic manifestations, over a given time at a given place. Within this definition, both a very broad and abstract concept and also precise descriptions and concrete representations are conceivable, depending on the scale and extent to which a specific soundscape is examined in its temporal and spatial dimensions. Accordingly, I introduce two different planes of analysis: the microsoundscape and the macrosoundscape.

A microsoundscape contains all acoustic elements (relevant for human perception) present within a defined time segment at a specific place. A microsoundscape is always a concrete, real soundscape that can – at least in theory – be completely and entirely described by listing all of its existing elements. It can be recorded and reproduced (given suitable technological resources) in its exact composition and fairly complete scope, depending on the chosen resolution and technical limitations. A concert in a hall is one example of a microsoundscape. A millisecond of silence in an acoustic anechoic chamber would be another. A macrosoundscape, on the other hand, is conceptualised as the sum of all existing (concrete) microsoundscapes at a specific place and time that cannot be represented in a single concrete microsoundscape due to the particular spatial and/or temporal extensiveness. A full macrosoundscape contains all audible information at a specific place and time, represented as a set of all possible microsoundscapes of the specified place at the chosen time (elements). While the microsoundscape is a concrete audible entity, the macrosoundscape is an abstract concept which allows us to contemplate such soundscapes as the year 1945 as a whole in an abstract manner. Also, it enables us to examine soundscapes as concepts, the analysis of which does not require the precise definition of the spatial boundaries, the exact timeframe or all of the concrete microsoundscapes contained within it. Hereafter, whenever clarity demands it, I will refer to the exact soundscape concept meant.

My working definition of soundscape research is a broad one that could be defined as any systematic effort to study phenomena within the above-mentioned concept of the soundscape, as distinct from the much narrower notion of the study of soundscape being primarily the subject of acoustic ecology or soundscape ecology.[1]

---

1 Soundscape research in a narrower sense, understood as acoustic ecology, has been significantly influenced by authors such as R. Murray Schaefer, Bernie Krause, Hildegard Westerkamp and Barry Truax.

The muteness of war trauma and its narrative context

The beginning of the war is easy to pinpoint, but its end is difficult to determine.[2]

During the twentieth century, a great amount of research and literature was devoted to the examination, description, understanding and theorisation of the circumstances surrounding the catastrophes of war.[3] The rise of psychoanalysis at the beginning of the last century affected the humanities and social sciences in various ways, leading to the rapid emergence of new theories and interdisciplinary fields. However, its rich influence is predominantly found in theoretical works, while practical applications have for the most part been limited to psychological areas, most prominently psychotherapy. When approaching the topic of war tragedies and the traumas related to them, in particular after the end of the Second World War, a division into two spheres of action can be discerned: collective (public) and individual (private). At the same time, the theory-practice dichotomy has determined the separation between collective and individual focus on the traumatic past, as well as the understanding of its constituent elements, impact, consequences for subjects and potential for recovery. In the immediate post-war times, following the year 1945, the main focus of efforts to deal with the traumatising past appeared to be placed in the broader sociopolitical context, the impact of which was witnessed in everyday life, affected by the different society-shaping ideologies that were implemented after the war. The life of the individual was substantially influenced by the political decisions enforced in the radically split East and West of post-war Europe. The reconfiguration of the world and the rebuilding of society, institutions and identities was a political and above all economic matter. Whatever ideology fuelled the sociopolitical reality of a particular post-war society, individual resources were expected to be put first and foremost into the function of the collective remediation of the destroyed land, rebuilding and strengthening the new identity and economy.

The described utilitarian aims of the immediate post-war period in devastated Europe were focused on building a new foundation without direct reference to the traumatic past, providing distance and a clear separation from the war while avoiding open wounds. The main political focus on stabilising the sociopolitical conditions of post-war existence, largely defined by a characteristic ideological blindness to reality, engendered negligence over acknowledging the severity

2 Orwid and Bomba, "A psychiatric study", 220.

3 Different perspectives and approaches focusing on the effects of these experiences on individuals and society are assembled in the broader trans-disciplinary field of trauma studies.

of traumas and a disregard of the importance of addressing them in the public discourse, which was perhaps not always a specific or declared strategy for subordinating individual lives and needs. As a consequence, traumatic personal experiences and memories of raging violence, constant threat, displacement and a general state of hopelessness, as well as a lack of control over one's own life, were omitted from the collective efforts and not addressed as a substantial, recurring issue within the public coping discourse, not even among those who articulated them in numerous testimonies. Life was to "return to normal", although the construct of normal had yet to be redefined, after its former definition had been blackened in the vocabulary of many who witnessed the appalling devastation.

As a result of the dominance of such public discourses, creating a division into the private and the public spheres, the personal aspect was virtually silenced. The phenomenon of untold personal stories is present among the war generations and their offspring regardless of the position they inhabit in the war discourse. For example, Lawrence Powell's book about Anne Skorecki Levy, a Polish Holocaust survivor, depicts a dramatic family history characterised by decades of silence originating from the pre-war and war-time experiences of Jewish survivors. It illustrates the effects those experiences and coping realities had on the lives they subsequently made for themselves.[4] Refusing to talk about their war experiences and personal trauma, even privately or to a therapist, was also a frequent response among Holocaust survivors living in Poland and their children, evident in the psychiatric studies conducted by Polish psychiatrist and family therapist Maria Orwid and her team.[5] From the opposite perspective, cultural studies scholar Aleida Assmann identified keeping silent about individual experiences from the war as the predominant common coping strategy related to the traumas resulting from the Second World War within families on both sides of post-war Germany.[6] Furthermore, sociologist and peace studies researcher Ivana Milojević analyses the mechanisms of violence, war conflicts and their consequences in the context of both collective history and individual biographies from a post-soviet perspective. She indicates that keeping silent about the past and the stories containing unspoken memories of her ancestors contributed to the accumulation of different layers of trauma passed on through the generations, leading to serial breakdowns in personal biographies.[7]

These biographical accounts, studies and theoretical explorations testify, albeit exemplarily, that in the period following the war tragedies, a narrative vacuum

4 Powell, *Troubled Memory*.
5 Orwid and Bomba, "A psychiatric study".
6 Assmann, *Das neue Unbehagen*.
7 Milojević, *Breathing*.

often emerged in the individual sphere, as a result of accumulated traumas followed by various destabilising circumstances, such as endangered existence through migration and forced resettlement, fear of repeated or further repression, persecution or prosecution, loss of identity and close social connections, inability to communicate due to the language barrier, inner struggles with the question of one's own guilt, and the general political discourse of the time. In such surroundings, defined by a reality of life in which every one of the aforementioned factors bears the risk of further or repeated re-traumatisation, troubled memories were naturally suppressed and locked down into an archive of private memories that continued to exist in the minds of their witnesses but were otherwise revoked from the explicit narratives of personal biographies. The narrative of the trauma was adapted to the social context and fell silent in reaction to the lack of stability and empathic climate in an environment indicating that personal stories carried little or no relevance in the tough times of reconstruction. This social climate was compounded by the fear of being misjudged and the inability to discern a meaningful purpose that might be fulfilled by telling the story within the given public discourse.[8]

> Stories are not told in a vacuum; rather, they are told within social settings and embedded within social actions.[9] As a result, stories are told with the audience in mind and storytelling is participatory (even if it is passive and limited to backchannel cues).[10] Stories are also situated within the history and culture of the moment.[11]

Bearing in mind that the political discourse of the first five decades of the twentieth century was largely dominated by the hegemony of nationalistic and violence-advocating forces aiming to defer their population to the "higher" interests of the nation, the oppressive culture of the time promptly rendered any individualist and non-conformist positions as disloyal and treacherous, in many cases to fatal effect. The generations that grew up in such reality were socialised within a public discourse which made them feel that the individual was of little – if any – account and was therefore likely to face inner restraints when addressing their own personal traumas later in life. Change only became possible when the next generation was socialised in the post-war society. So it was the offspring, the so-called Generation

---

8 The following passage is quoted, with the original notes, from Jackl, "Rules of telling", 3.

9 Gergen and Gergen, "Narrative and the self"; Langellier and Peterson, *Storytelling*; Langellier and Peterson, "Narrative performance theory"; Ochs and Capps, *Living Narrative*.

10 Langellier and Peterson, "Shifting contexts"; Langellier and Peterson, *Storytelling*.

11 Langellier and Peterson, "Narrative performance theory".

of '68, that would challenge the discourse by distrusting the still prevalent concept of not dealing with the consequences of history, along with the military elites and authorities including their parents, and demonstratively rejecting the repressions imposed by institutions, rigid politics and a morally paralysed society.

The phenomenon of the unspeakable in the context of trauma can be further approached from the direction of narrative theory, showing that the muteness of trauma can be reinforced by obstacles found in the narrative of the story itself. Narratology "examines the ways that narrative structures our perception of both cultural artifacts and the world around us".[12] It has developed into a group of related theories, incorporating a variety of analytical procedures.[13] "Its concepts and models are widely used as heuristic tools, and narratological theorems play a central role in the exploration and modelling of our ability to produce and process narratives in a multitude of forms, media, contexts, and communicative practices".[14] Narratology studies the narrative – a description of a series of events, real or imagined, often told in written or spoken words, still or moving images, or sometimes expressed through other media – and the narrative structure, being the system that underlies the way in which a narrative is presented to a reader, listener or viewer. Thus from the perspective of narrative theory, telling stories is a process that can be seen as preconditioned by the existence of content that should be communicated, and by the means of communication. So there are two angles that can be taken in examining the constraints around the act of telling traumatic stories: the narrativity of a traumatic situation, and its tellability.

In his text on narrative identity, Wolfgang Kraus addresses several possible disturbing elements affecting narrativity in stories. Three factors can be especially destructive to the narrativity of war traumas: the presence of (1) an event without a concrete opponent, (2) extremes in the continuance in time, and (3) extremes in the representation of causality.[15] In the context of traumatic experiences, a war can feel as if (1) there was no concrete opponent but only abstract opponents such as human destructive nature. The fact that over four decades Europe was devastated by the extensive violence of the two world wars, the ongoing and never-ending insecurity of the times and the fear of a repeat of the horrors already brought on people by the First World War in the years following it and preceding the Second World War constitute (2) an extreme in the continuance in time. Extremes in the

12 Felluga, "General introduction".
13 Meister, "Narratology" (2014).
14 Meister, "Narratology" (2009), 329; *Narratology – the Living Handbook of Narratology.*
15 Kraus, "Arbeit am Unerzählbaren".

representation of causality (3) can occur within narratives of war traumas in both directions. On one hand, it might seem impossible to find any coherent causal chain that would explain either the reasons for the inconceivable extent of the war and the brutality of global catastrophes or the extreme personal experiences that happened. On the other hand, one might identify numerous events and precursors that, taken all together in very complex interaction with each other, were all paving the way to a horrible end. However, such complicated stories reduce the narrativity. They overcomplicate the story, burden it with too many interwoven elements influencing the overall situation, and so cannot be completely understood and properly performed or perceived.

A story characterised by a negatively affected narrativity resulting from any of these factors can evoke unpredictable reactions from the interaction partner due to intellectual or emotional overload. A consequent negative reaction can therefore place a further burden on the already unbearable, unspeakable experience and defer any further attempts to communicate the trauma.[16] For some, an assumed negative experience from such a situation can already strengthen the reluctance to expose one's own trauma in the first place, suppressing the whole narrative.

The other direction to examining the obstacles to relating extreme experiences from the position of narrative theory points to the concept of the tellability of the story. Elinor Ochs and Lisa Capps analyse tellability with the focus on the reasons for which a narrative is told.[17] Most narratives are tellable because they are different from everyday occurrences and prototypical stories. However, the tellability of a narrative is dependent on the content, the narrator and the audience, and it is strongly influenced by a concern about what is allowed to be told. The question of the content and its influence on the narrative process was discussed above within the context of narrativity. A person's willingness to share stories with others is substantially shaped in early childhood through socialisation within the family and absorbed responses from the everyday social setting.[18] This means that growing up in an environment characterised by uncertainty, increased cautiousness and fear can result in enduring internalised censorship. Besides upbringing, the assessment of sociopolitical circumstances, personal stability and other existential issues are important factors in deciding if and when an experience can be shared with others, and with whom specifically. After all, a story that nobody wants to hear is difficult to tell.

16 Ibid.

17 Ochs and Capps, *Living Narrative.*

18 Ibid.

However, the concern over what is allowed to be told should not be understood in terms of the assessment of the risks imposed by external factors and feared sanctions alone. It is rather the underlying cause of the act of self-censorship that is commonly seen as a strategy for dealing with trauma. It represents a consciously chosen path, often described as a decision made to "protect the children" from the awful past.[19]

> What results, however, is a "conspiracy of silence",[20] a psychological defense mechanism aimed at the prevention of intrusive traumatic memories and emotions and at the continuation of life-family cycles as if "nothing had happened". Further, survivors are often faced with the following dilemma: acknowledging the damage done may be seen as the ultimate victory by the perpetrators of violence; not acknowledging it, on the other hand, may seem as the ultimate betrayal of the victims who perished.[21] Conflicted with this, and told to "let bygones be bygones",[22] survivors often elect to not talk about their experiences.

Finally, according to Ernst van Alphen, the key question in determining the ability to communicate about extreme experiences is whether the experience has been narratively processed at all. Van Alphen argues that an event becomes an experience only if it can be framed within an existing discourse. However, in the case of trauma, it becomes a "failed experience", an event that cannot be processed with adequate forms of representation and cannot be incorporated into one's own history because of the inability to make sense of the extreme event. "Experience does not really exist until it can be named and placed in larger categories".[23]

The above discussion presents arguments made from the perspective of historical trauma to demonstrate the possibility of approaching the topic of the muteness of trauma. It shows that the reasons for keeping silent about extreme personal stories are manifold and found in both private and public discourses, continuously inhibiting the capacity to speak about experienced trauma. The resultant exclusion of a significant portion of war memories from the narratives of post-war existence genuinely influenced social interactions and behaviour, which correspondingly found its manifestation within the soundscape. Examined from this perspective, the (macro)soundscape of 1945 and the subsequent period can be understood as characterised by the systematic absence of the verbal representation of silenced traumatic memories. That absence of verbally articulated accounts of individual

---

19 The following passage is quoted, with the original notes, from Milojević, *Breathing*, 237.

20 Danieli, *International Handbook*; Danieli, "Recovery".

21 Danieli, *International Handbook*.

22 Ibid., 4.

23 Kolk, McFarlane and Weisaeth, *Traumatic Stress*, 488.

traumas, which were a constitutive part of society in the wake of the Second World War, denotes war trauma in the auditory sphere as a symptom of the traumatising effects on the individuals. From this perspective, the examination of the muteness related to war trauma may be seen as a zoom-in on the social soundscape, observing it directly at its source, namely, individual articulation. In this approach, a link between trauma studies and soundscape research can be established by exploring connections between the concepts of trauma and the concerns and potential of soundscape research. Hence the focus of research is shifted from the audible surface and the description of the acoustic reality towards identifying and understanding the underlying mechanisms necessary for approaching unarticulated traces of the past in the auditory sphere.

Limits of verbalisation from the perspective of post-traumatic stress disorder

As argued from the narrative perspective, the prerequisites for the vocalisation of traumatic experiences are complexly interwoven with sociopsychological mechanisms determined by both individual and collective contexts from the past. Many texts from the field of narratology extensively address the issue of traumatic experiences, describing models within the linguistic system that are involved in the process of translating a traumatic event into experience and finding its representation in one's own life story. In so doing, the authors are at the same time involved in the construction of models for trauma processing, offering a theoretical foundation and an interface for linking with psychology, especially psychotraumatology, directed towards the development of practical applications for managing trauma. Conventional instruments for coping with trauma are being developed in psychotherapy and psychotraumatology in the form of methods and therapies traditionally based on a theory and/or experimental research. Even though the history of humankind is paved with countless tragedies, trauma studies, with its focus on understanding the mechanisms, causes, forms and risk factors of trauma, as well as the ways of managing and overcoming trauma, is a fairly young field of research. In my investigation of trauma-related verbalisation issues, I have indicated the link between the verbal-narrative origins of the predominant psychotherapies for traumatised persons and the dissatisfying results of these trauma-focused models in practical applications.[24] In this critical investigation, I have argued that the deficient effectivity of the widely-applied therapies for PTSD (posttraumatic stress

24 Lange, "Going beyond words".

disorder) can be identified as a consequence of the fixation on the linguistic and verbalisation concepts in their therapeutic models. A significant proportion of traumatic experience and potential for healing remains beyond the reach of the employed verbal methods, as is evident from the disappointing success rates of the evaluated trauma-focused cognitive-behavioural therapies.

By pointing out that psychotherapy and academic research operate on the basis of a common underlying verbal cognitive-narrative logic, I argue that soundscape research focused on the description, analysis and discussion of acoustic qualities relies in its essence on narratives operating in verbal modes, depending on language as its medium of reasoning.[25] One of the crucial insights from this standpoint is that it can consequently be implied that research into trauma-laden soundscapes conducted within the bounds of established academic discussions utilising verbal modes of reasoning cannot provide the full framework necessary to effectively address the traumatic context, in either theory or practice. As a result, the outcome of soundscape research based on verbal methods remains – in terms of effective reflection on the traumatic nature behind the subject matter and capacities for illuminating those aspects of the reality that have been suppressed – constrained as well. Especially in the context of extreme experiences, given the obstacles surrounding the narrative of trauma and the frequently reported difficulties with expressing trauma in words, it seems obvious that accessing the muted past through alternative models should be considered with more attention paid to their nonverbal potential, supplementing the commonly applied language-driven approach. Moreover, by revealing the physiological mechanisms behind the phenomenon of inhibited abilities to actively attend to traumatic experiences through verbal cognitive modes, the result of neuroscience studies involving traumatised persons strongly support the arguments for a necessary nonverbal shift from yet another perspective.

Approached from this position, it becomes clear that studying soundscapes in search of representations of individual traumas, that is, traces of traumatic experiences in the social reality of the post-war times, should not be limited to forms of verbal articulation. Moreover, this means that a significant proportion of post-traumatic individual existences (and therefore social reality) lie beyond the narrow band of verbal expression, which can act as a selective silence masking the trauma. It is therefore necessary to critically assess approaches that are limited to seeking evidence of trauma and traumatic experiences in verbal narratives, since they can also be found in nonverbal artefacts of those forms of expression that

25 Ibid.

lie beyond verbalised biographical and historical accounts. The investigation of the requisite means for a successful revelation of the respective nonverbal and verbally inaccessible traces of trauma through the medium of sound can therefore be postulated as one of the concerns of the nonverbal perspective in approaching the soundscapes of traumatic environments. Such an approach to studying acoustic phenomena as a form of trauma articulation, together with their "laws", that is, the principles behind their emergence and materialisation, could also be understood as offering the possibility of including soundscapes in the broader discourse of trauma studies.

Two examples of artistic nonverbal concepts between war trauma and sound(scape)

The following examples demonstrate different artistic concepts for researching war trauma in nonverbal modes with the soundscape as their working medium. The first example, a sound and video installation, denotes a concept of nonverbal representation of trauma in the soundscape. The second, a stage performance, expands the focus of the nonverbal research possibilities beyond the representation issue of the trauma. It illustrates an artistic research approach orientated towards the investigation of nonverbal potentials for transforming and overcoming trauma through the medium of sound.

Representation of (in)audible war trauma in a soundscape

The sound and video installation *9-11-1938* by Boris Hegenbart and Volker Straebel, commissioned to commemorate the seventieth anniversary of the *Reichspogromnacht* on 9 November 1938, was exhibited from 9 November to 7 December 2008 at the Collegium Hungaricum in Berlin, Germany. It occupies a place within the tradition of memorials dealing with German history and the consequences of the twelve years of Nazi terror. The essence of the installation form in this category is that the concept of public space sound art is transferred to the format of a memorial. The artists described their involvement as follows:

> Based on the belief that the persecution of the Jews and the Holocaust cannot be represented in narrative accounts, and that an atmospheric or illustrative approach to the pogrom of November 9th, 1938 results in the trivialization of this genocide and injures the dignity of its victims, we have created a sound installation which conceptually focuses on empty space, that the extermination of Jews and other groups has left in the life of European nations.[26]

---

26 Hegenbart and Straebel, "9-11-1938"; Straebel, "Volker Straebel".

Hegenbart and Straebel's work is situated in the field of sound art, between music and the visual arts, and emerges from a special interpretation of the emptiness and silence in their approach. Utilising artistic means, they produce an aesthetic situation in which silence, the absence of sounds and of sight, becomes apparent.[27] Their video projection, which could be seen from the street as part of the Collegium Hungaricum building's façade, visually removes a part of the reality, erasing it as if it did not exist, by creating the illusion of looking through the house.

> On the levels of the second and third floors on the south façade of the Collegium Hungaricum Berlin there is a large panoramic window which can be used as a video screen via a reverse projection. Here a live video signal will be transmitted, which comes from a camera, which is directed out of a window on the north face of the building, towards a row of houses on Bauhofstrasse. This produces a visual opening on the south façade, which enables the viewer to look through and into the building.[28]

The acoustic realisation of the installation was located inside the building, where the artists reproduced in real time the soundscape of the street in front of the building. As the source material for their sound installation, they employed the profusion of this seemingly trivial everyday reality, which we are used to ignoring in our everyday lives. Similar to the concept behind the visual idea used on the outside walls of the Collegium, a filtering process that deletes a narrow acoustic band from the original full-spectrum soundscape of the local outside reality was applied to the sound material. This process makes the otherwise hidden reduction, extraction, deficiency and absence acoustically evident. Nevertheless, it remains a highly delicate process demanding the utmost concentration and dedication of the observer in order to perceive and comprehend what has been erased and what is consequently absent. With this concept, Hegenbart and Straebel succeeded in making emptiness and silence audible and visible. In their installation, the missing parts that were consistently filtered out of the everyday sounds stand for the loss that the extermination of the Jews, a permanently erased part of the urban soundscape, denotes for German culture. Visitors to the installation who could recall these occurrences from the distant past would become aware of the fact that the consequences of the Nazi terror were still noticeable even seventy years later and that it is still impossible to understand the present without knowing the past. However, Hegenbart and Straebel's work is not about an abstract reconstruction of German history, but rather about the perceiver's self-assurance about the historical place in which they find themselves.[29]

---

27 Sanio, "Die Paradoxie".

28 Hegenbart and Straebel, "9-11-1938"; Straebel, "Volker Straebel".

29 Sanio, "Die Paradoxie".

Exposing trauma by dissecting the soundscapes of distressing memories

The second example, *Trans(ar)chiving*, is my own artistic research project being developed as a final thesis project within the Sound in New Media programme at Aalto University in Finland. Its final outcome is conceived as a thirty-minute stage performance incorporating video projections, musical compositions and improvisations for flute, piano and model aeroplane, along with the real-time manipulation of sound. The NATO bombing of Serbia in the spring of 1999, a formative trauma from recent European history, defines the thematic background of the project. Within the discourse of the artistic research, the possibilities of actively transforming and overcoming distressing memories through an alternative narrative of the experienced traumatic event are being investigated. Within that process, sound and new media play key roles as artistic elements whose potential and impact should be explored. The artistic research model developed in the project can be described by naming its four stages: (1) accessing a traumatic experience from the past through exposure to reminders of that experience while working with various (documentary) media such as video, audio and text; (2) understanding the inner reactions to the reminders of trauma through extensive research, and reflecting on the hurdles to the exposition of the trauma; (3) achieving self-empowerment through the regaining of control over the disturbing traces of the past by identifying appropriate liberating channels for expression and free experimentation; (4) performing *trans(ar)chiving*, an attempt to transform the old, trauma-contaminated narratives into new ones. This is a performative act that avoids accessing traces of the traumatic past in an expected way, such as trying to express them through a verbal narrative in the "authentic" form of the traumatic incident as it was experienced, or as it is still remembered. The disturbing recollections, such as the hopeless situation of being condemned to waiting for the disaster to be over, or the actual sounds of the circling aeroplanes and explosions, are not reproduced. Instead, these reminders, in the form of documentary material, told stories and inner recollections of the original traumatising situation, are paired with composed sounds of a quality essentially different from the original disturbing ones. In a concrete example, during the performance projection of the original video scenes showing the missiles falling and exploding on the ground, the authentic sound of the event is removed and replaced by a composition with an aesthetic of underwater buffered sounds that evoke a serene feeling of a safe distance, far away from the battlefields on the ground. Through the continuous imposition of these procedures on different disturbing elements of the initial trauma, a new performative narrative is established, one which should co-exist with the original narrative that stands for the inaccessible experience causing emotional overload, suffering or other difficulties. Instead of attempting an expressive materialisation of the experienced past, the idea behind the method is aimed rather at an effective remodelling of the verbally inaccessible traces into an alternative nonverbal form that can be used to work with and build upon. From the soundscape perspective, the work can be described as an attempt at dismantling the remembered original, destructive soundscape of the trauma and replacing it with a newly composed one that has been developed during intense (artistic) engagement on the nonverbal level. In this concept, sound is declared to be a key coordinating element. As such, it is hypothesised as being capable of generating an impact on the nonverbal cognitive level significant enough to support the reorganisation of behavioural patterns. The project is therefore an experiment integrating approaches from trauma studies, psychotherapy, neuroscience, sound studies, performance studies and art therapy. The outcome of the experiment should provide insights into the potential of sound and soundscape as working media for sublating trauma beyond merely making hidden traces of the past audible.

Conclusion

By acknowledging that massive war traumas are constitutive elements of every post-war society, it becomes obvious that the study of the soundscape of such environments needs to be conducted with sufficient sensitivity to the described verbalisation limits. This requires new concepts and methods, accommodating and adapting nonverbal approaches, which can be developed beyond disciplinary boundaries. Soundscape research needs to reflect on its own nonverbal potentials in order to work out and emphasise the significant advantages that it can claim for itself in distinction from the discourses found in other approaches that operate exclusively within verbal boundaries. Gaining further insight into the (social) reality of post-war society, one of the principal goals of the attempt to study the soundscape of 1945 can be achieved at the intersection of the soundscape and other cognitive domains, and it should be understood as a sum of both verbal and nonverbal spheres. Trauma studies and artistic practices can offer useful tools for accessing and revealing the remains of the past within soundscapes. Some closing impulses for studying soundscapes between war and trauma could perhaps be formulated as questions needing further attention.

What are the concrete nonverbal manifestations of trauma in the soundscape of post-war times and how do/did these forms of articulation sound? How can they be approached, recognised and made legible? Do these artefacts and hidden traces of trauma reflect massive war losses? And how can a lack of verbal manifestations in the reality of 1945 and afterwards be conceptualised and represented?

What role can soundscape research play in examining the traumatic contexts of individual or collective suffering that other disciplines cannot offer to the same extent? What would be the potentials that soundscape studies could offer as a research practice focusing on the reflection of the medium itself? And, if it can be asserted that there are traumatising soundscapes capable of transporting traumas and causing harm, could there be trauma healing soundscapes as well?

Bibliography

Danieli, Yael, *International Handbook of Multigenerational Legacies of Trauma* (New York: Plenum Press, 1998).

– "Recovery after mass crimes", in *The Oxford International Encyclopedia of Peace*, iii: *Mead, Margaret – Rwanda*, ed. Nigel J. Young (Oxford: Oxford University Press, 2010), 615–18.

Felluga, Dino, "General introduction to narratology", https://www.cla.purdue.edu/english/theory/narratology/modules/introduction.html, accessed 1 November 2017.

Gergen, Kenneth J., and Gergen, Mary M., "Narrative and the self as relationship", *Advances in Experimental Social Psychology*, 21 (1988), 17–56, https://doi.org/10.1016/S0065-2601(08)60223-3.

Hegenbart, Boris, and Straebel, Volker, "9-11-1938", http://www.soundblocks.de/sound-video-installtion-09-11-07-12-2008/, accessed 1 November 2017.

Jackl, Jennifer A., "Rules of telling: Exploring narrative adaptation", *Journal of Social and Personal Relationships*, 30 November 2016, 0265407516681539. https://doi.org/10.1177/0265407516681539.

Kolk, Bessel A. van der, McFarlane, Alexander C., and Weisaeth, Lars (eds), *Traumatic Stress: The Effects of Overwhelming Experience on Mind, Body, and Society* (New York: Guilford Press, 1996).

Kraus, Wolfgang, "Arbeit am Unerzählbaren. Narrative Identität und die Nachtseite der Erzählbarkeit", in Elisabeth Waller, Anja Stukenbrock, Gabriele Lucius-Hoehne and Carl E. Scheidt (eds), *Narrative Bewältigung von Trauma und Verlust* (Stuttgart: Schattauer, 2015), 109–20.

Lange, Dorian, "Going beyond words: Considerations for approaching trauma in soundscapes", *Soundscape. Concepts-Research-Practice. The Polish Soundscape Journal*, 1 (2017), http://pracownia.audiosfery.uni.wroc.pl/numer-152017/.

Langellier, Kristin M., and Peterson, Eric E., "Narrative performance theory: Telling stories, doing family", in Leslie A. Baxter and Dawn O. Braithwaite (eds),

Engaging Theories in Family Communication: Multiple Perspectives Engaging Theories in Family Communication: Multiple Perspectives (Thousand Oaks, CA: SAGE, 2006), 99–114, https://doi.org/10.4135/9781452204420.n7.

– "Shifting contexts in personal narrative performance", in D. Soyini Madison and Judith Hamera (eds), *The SAGE Handbook of Performance Studies* (Thousand Oaks, CA: SAGE, 2005), 151–68.

– *Storytelling in Daily Life: Performing Narrative* (Philadelphia: Temple University Press, 2004).

Meister, Jan Christoph, "Narratology", in Peter Hühn, John Pier, Wolf Schmid and Jörg Schönert (eds), *Handbook of Narratology* (Berlin: De Gruyter, 2009), 329–50, https://doi.org/10.1515/9783110217445.

– "Narratology", in Peter Hühn, Jan Christoph Meister, John Pier and Wolf Schmid (eds), *Handbook of Narratology*, 2nd rev. edn, (Berlin: De Gruyter, 2014), 623–45, https://doi.org/10.1515/9783110316469.

Milojević, Ivana, *Breathing: Violence In, Peace Out* (Brisbane: University of Queensland Press, 2013).

"Narratology – the Living Handbook of Narratology", http://wikis.sub.uni-hamburg.de/lhn/index.php/Narratology, accessed 1 November 2017.

Ochs, Elinor, and Capps, Lisa, *Living Narrative: Creating Lives in Everyday Storytelling* (Cambridge, MA: Harvard University Press, 2001).

Orwid, Maria, and Bomba, Jacek, "A psychiatric study of World War II survivors: The case of Poland", in Jolande Withuis and Annet Mooij (eds), *The Politics of War Trauma: The Aftermath of World War II in Eleven European Countries*, Studies of the Netherlands Institute for War Documentation, 4 (Amsterdam: Aksant, 2010), 217–239.

Powell, Lawrence N., *Troubled Memory: Anne Levy, the Holocaust, and David Duke's Louisiana* (Chapel Hill, NC: University of North Carolina Press, 2000).

Sanio, Sabine, "Die Paradoxie des Leeren", in János Can Togay and Bettina Wackernagel (eds), *09-11-38* (Berlin: Collegium Hungaricum, 2008), 12–13.

Straebel, Volker, "Volker Straebel: Thoughts on music and related arts/Musikwissenschaftliche Praxis", http://www.straebel.de/praxis/index.html?/praxis/text/t-9-11-1938e.htm, accessed 1 November 2017.

Authors

Dorota Błaszczak PhD is an audio engineer at the Polish Radio Archives (PRA) in Warsaw and a lecturer on the Sound Engineering Department of the Fryderyk Chopin University of Music in Warsaw. She leads courses in interactive sound, works in sound restoration and preservation and concert sound projection, participates in the development of the PRA Multimedia Database, and creates interactive projects (e.g. *Hearing Data. Data Carriers* (2014), *Radio Receiver of 90 Years* (2015), *Waves of Remembrance: Wrocław in Radio Sounds* (2016)). She has been sound designer for VR (virtual reality) projects in Canada, including at the Banff Centre for the Arts, Softimage and Immersence (e.g. Lawrence Paul Yuxweluptun's *Inherent Rights, Vision Rights* (1992), Perry Hoberman's *Bar Code Hotel* (1993), Char Davies's *Osmose* (1995) and *Ephemere* (1998, remastered 2017)). She also designed the sound for some early computer games.

Daniel Brożek – a Wrocław-based sound art curator (Survival Art Review) and modern music critic (*Glissando* magazine, Canti Illuminati blog). In the Czarny Latawiec project, he works as a sound artist and produces music albums, sound installations and incidental music for plays.

Andrzej Dębski is a film historian at Wrocław University's Willy Brandt Center for German and European Studies, where he is conducting the research project "History of cinema in Lower Silesia until 1945". He was awarded his doctorate at the University of Silesia in Katowice for *Historia kina we Wrocławiu w latach 1896–1918* [The history of cinema in Wrocław 1896–1918], published in 2009. He co-edited *Wrocław będzie miastem filmowym… Z dziejów kina w stolicy Dolnego Śląska* [Wroclaw will be a film city… The history of cinema in the capital of Lower Silesia] (2008, with Marek Zybura), *W drodze do sąsiada. Polsko-niemieckie spotkania filmowe* [Approaching the neighbours. Polish-German cinematographic contacts] (2013, with Andrzej Gwóźdź; German version *Unterwegs zum Nachbarn. Deutsch-polnische Filmbegegnungen*, 2015, with Brigitte Braun and Andrzej Gwóźdź), *KINtop. Antologia wczesnego kina* [KINtop. Anthology of early cinema] (2016, with Martin Loiperdinger), *Stanisław Lenartowicz* (2011) and *Sylwester Chęciński* (2015; both with Rafał Bubnicki).

Annelies Jacobs is a lecturer at the Faculty of Arts and Social Sciences of Maastricht University (NL). She received her PhD in Science, Technology and Society

Studies from Maastricht University and has also been trained as an architect at Eindhoven Technical University. Her most recent publication was "The silence of Amsterdam before and during World War II: Ecology, semiotics and politics of urban sound" (2014), in Daniel Morat (ed.), *Sounds of Modern History. Auditory Cultures in 19th and 20th Century Europe* (New York: Berghahn Books).

Karolina Jara is a doctoral candidate at the Institute of Art History of University of Wrocław. She is preparing a PhD thesis on architecture and urban planning in Silesia during the period of National Socialism under the supervision of Associate Professor Agnieszka Zabłocka-Kos. She received her MA in Art History from University of Wrocław in 2013 with a thesis on the New Synagogue in Wrocław (1865–1872) and its designer Edwin Oppler (1831–1880). Her main fields of interest are architecture, urban planning and design from the nineteenth century to the present day, especially the Polish-German heritage. She also works with the Transformator art foundation, which aims to promote and popularise knowledge about architecture and urban planning.

Dorian Lange holds a degree in Computer Science and Digital Media from the University of Bremen with a focus on digital media in education and experimental methods in artistic research. He has since focused on working within the context of media events, exhibiting and curating media art within the MFA program in Media Art and Design at the Bauhaus University Weimar. At present, he is conducting a combined artistic/research thesis in Sound in New Media at Aalto University, Finland. He is a fellow of the Hans Böckler Foundation.

Katarzyna Naliwajek-Mazurek writes on Polish contemporary music, opera, aesthetics, and the relationship between music and politics at the turn of the 1940s. A recipient of the Hosenfeld/Szpilman Gedenkprize at the University of Lüneburg (2011) for her multimedia exhibition *Music in Occupied Poland 1939–1945*, she also received the Literary Prize of the City of Warsaw and the "Klio" Historical Prize for the first of two volumes of *Okupacyjne losy muzyków, Warszawa 1939–1945* [Musicians in occupied Warsaw 1939–1945], co-authored with Elżbieta Markowska (2015). She is a board member of the Witold Lutosławski Society and a member of the repertoire committee of the "Warsaw Autumn" International Festival of Contemporary Music.

Zoriana Rybchynska – PhD, Head of the Cultural Studies Department at the Ukrainian Catholic University (Lviv). Ukrainian modernist literature studies and its close ties with the spaces of the modern city inspired her research interest in cultural landscapes and in the relations between language, place and the people who inhabit it.

Uta C. Schmidt studied historical science and art history at the Ruhr-Universität Bochum; she gained her PhD at the University of Bielefeld. In her research, she is particularly interested in topics dealing with space, representation, gender and power. She works as a public historian and develops projects dealing with historical culture and contemporary witnesses. She is currently co-manager of the 7th Historical Competition "HAU REIN! Bergbau im Ruhrgebiet. Alltag. Wissen. Wandel". In 2017, she became a research associate at the Women's and Gender Research Network NRW. Since the early 1980s, together with Richard Ortmann, she has helped to document the soundscape of the Ruhr, and she was among the first scholars to highlight the connection between sound and historical change based on the example of the Ruhr.

Kamila Staśko-Mazur – PhD candidate at the Institute of Musicology of the University of Wrocław, member of the Soundscape Research Studio, author and participant of sound projects on acoustic ecology.

Renata Tańczuk is a professor of Cultural Studies at the University of Wrocław. Her main research areas are material culture and collecting, as well as sound and soundscape studies. She is a member of the Laboratory of the Contemporary Humanities, the Soundscape Research Studio and the editorial board of the *Polish Soundscape Journal. Concepts-Research-Practice*. Her publications include *Ars colligendi, kolekcjonowanie jako forma aktywności kulturalnej* [Ars colligendi: collecting as a form of cultural activity] (Wrocław 2011) and "The horror of natural history collections" (http://pkult.wuwr.pl/preview/4380, Wrocław 2012). She is also co-author and co-editor of several books, including *Audiosfera miasta* [The Urban Soundscape] (Wrocław 2012), *Audiosfera Wrocławia* [The sounds of Wrocław] (Wrocław 2014) and *Audiosfera. Studia* [The soundscape: studies] (Wrocław 2016).

Sławomir Wieczorek graduated in cultural studies and musicology. He is a faculty member of the Institute of Musicology at the University of Wrocław. Author of the book *On the musical front: socialist realist discourse on music in Poland 1948–1955*, his interests focus on the history of twentieth-century music and soundscapes.

Jadwiga Zimpel is an adjunct professor at the Institute of Cultural Studies of Adam Mickiewicz University in Poznań. She conducts research in the field of cultural urban studies, urban anthropology and urban sociology, with particular emphasis on the issues of everyday life, sensory experience and place.

Eastern European Studies in Musicology

Edited by Maciej Gołąb

Vol. 1 Paweł Gancarczyk / Lenka Hlávková-Mráčková / Remigiusz Pośpiech (eds.): The Musical Culture of Silesia before 1742. New Contexts – New Perspectives. 2013.

Vol. 2 Laura Vasiliu/Florin Luchian/Loredana Iaţeşen/Diana-Beatrice Andron (eds.): Musical Romania and the Neighbouring Cultures. Traditions–Influences–Identities. Proceedings of the International Musicological Conference July 4-7, 2013, Iaşi (Romania). 2014.

Vol. 3 Barbara Przybyszewska-Jarmińska: Marcin Mielczewski and Music under the Patronage of the Polish Vasas. Translated by John Comber. 2014.

Vol. 4 Tomasz Jasiński: The Musical Rhetoric of the Polish Baroque. Translated by Wojciech Bońkowski. 2015.

Vol. 5 Bogusław Raba: Between Romanticism and Modernism. Ignacy Jan Paderewski´s Compositional Œuvre. Translated by John Comber. 2015.

Vol. 6 Maciej Gołąb: Musical Modernism in the Twentieth Century. Translated by Wojciech Bońkowski. 2015.

Vol. 7 Wojciech Bońkowski: Editions of Chopin's Works in the Nineteenth Century. Aspects of Reception History. 2016.

Vol. 8 Ivana Perković / Franco Fabbri (eds.): Musical Identities and European Perspective. An Interdisciplinary Approach. 2017.

Vol. 9 Bożena Muszkalska (ed.): The Kolbergs of Eastern Europe. 2018.

Vol. 10 Renata Tańczuk / Sławomir Wieczorek (eds.): Sounds of War and Peace. Soundscapes of European Cities in 1945. 2018.

www.peterlang.com

www.ingramcontent.com/pod-product-compliance
Ingram Content Group UK Ltd.
Pitfield, Milton Keynes, MK11 3LW, UK
UKHW021831100926
8556IPUK00004B/283

9 783631 753361